Culture
of an Italian
Region
The Marche,
Guzzini
and design

Augusto Morello

Culture of an Italian Region

The Marche, Guzzini and design

Electa

Editorial coordination
Piergiovanni Ceregioli,
iGuzzini Study and Research Centre

Archive studies by
iGuzzini Study and Research Centre

Graphic design
Pierluigi Cerri
and Carla Parodi

www.electaweb.it

Contents

Italy from day to day

Experimentation and pragmatism

Italy is a special country. So believe the Italians, when they declaim its history and inventiveness, but also when they criticise their own tendency to overemphasise its creative élan. They also acknowledge – to their regret – the planning abilities and organisational skills that seem to distinguish other European countries, such as Germany and Great Britain.
Its foreign admirers – of which there are many, as we know – and even its not always unbiased detractors say that Italy is special. And yet, reading about what people say about us – in the press, essays, books – suggests that, in the final analysis (disregarding all the preconceptions of others and our own about them), this country is considered a type of workshop; a workbench of experiences, which are often lateral, but almost always important to the attentive eye – maybe for later benefit. When asked to comment on the situation in Italy during an intense period of worker strikes, the head of Italy's largest corporation quipped that Italy "is a laboratory; we've always exported, so we'll export them too".
If the others do things that the Italians would never do, the Italians do things that the others have never done; nonetheless, in recent times the national press curiously reflects a need for reassurance that certain current projects have already been developed elsewhere. As if Lodovico il Moro had wondered whether Leonardo had worked for the French, who hosted him afterwards; or the Medici had first asked about Michelangelo in London, even though they saw him often; or Adriano Olivetti had requested information about his designers in New York, then exhibited their typewriters and sewing machines. People just didn't even think about it. *Benchmarking* – that new catchword for the modern company – is beneficial when (and only when) it is used to learn rather than paralyse. Particularly in the case of those who – like the Italians – do not suffer examples gladly, especially when they tend to become rules. One might object that this is a characteristic of many peoples; but it holds for only some. For the Italians this is 'more true', itself a typically Italian expression, dating back to well before the discovery of the increasingly-adopted *fuzzy logic*, which goes beyond the black and white to appreciate the greys. The Italians have always known about *fuzzy logic*; others have only recently theorised it. And this is the virtue of a lateral – not to be confused with a chance or chaotic – approach which denotes a different way of planning, of organisation. It is often the case that subsequently the others (as they themselves know) are the ones who verticalise, that is institutionalise, the experience of an organisation that here in Italy, as a rule, has nothing of a rule – even when the rules are written. Whatever they say, it is precisely in this that the Italians are a pragmatic people, leading indeed to a paradox: this country, notoriously famous for its aversion towards rules in the name of a more 'natural' pragmatism, is not considered pragmatic at all.
And yet its laws are amongst the most 'normative' (*batch* rules are hardly an Italian invention), which helps explain judgements that are often contradictory, the 'twin' strategic and operative level of the consciousness of events, culminating even in a schizoid state of affairs in the place of the fruitful paranoia of creativity. The fact is that rules exist – yes indeed – and many of them. But the ones that are most adopted, followed, enforced and enacted are not the written, but the verbal ones. In communication, they are often non-verbal, on the limit of consciousness. Not to mention the non-infrequent language of gestures, like those of the commedia dell'arte.
Steinberg noted as much in the Milan Galleria; in the style typical of this great

Saul Steinberg, *the Galleria Vittorio Emanuele, Milan.*

satirist and cartoonist, in the 1950s he depicted the Italians speaking with their hands, their arms, a tilt of the head, their proxemics sometimes considered vulgar or bad-mannered. The term *pernacchio* (Neapolitan version of the raspberry), made famous by the film featuring Eduardo, comes from the word *vernaculum*. To describe them, Steinberg realised that he had to look at people from above – therefore from the outside, the best way to capture them inside, thereby precluding any overlap of personal preconceptions.
But to recall (and, obviously, contradict at the same time) the famous non-definition made by Metternich ("Italy is only a geographical expression"), we must bear in mind that this is a country that has always known internal borders, and many at that. After the Romans and before the political unification of the country in the second half of the nineteenth century, Italy was never divided up into less than seven States – meaning a minimum of six and a maximum of twenty borderlines.
A multitude of different languages were spoken for almost a thousand years. Italian entered literature in the 11th century in Sicily, in the 12th century in Tuscany, from where it was to set the standard, though neither before or after in the French fashion, by means of an official scientific commission, the Academy. Italy also has known its share of academies, some with famous names: the Linceo, the Crusca, the Ligustica – but as usual (at least when they were operational), spontaneous. In certain Regions, like those in the north, the Italian language acquired official status much later – such as in Piedmont itself, from where the movement towards political unification initiated, appearing only in 1570. Quite late for the history of a culture. Besides the languages spoken in border areas, such as Ladino or the Slavic tongues, and those used in the southernmost Regions, like modern Greek or Albanian, dialects held the ground for centuries, right through to the Second World War; and indeed in certain Regions, dialects are still spoken in the family and even in the church, with accents that distinguish virtually adjacent towns from one another. Dialects are in fact an opportunity for local and official realities to come face to face. In the Emilia Region, things are said in dialect, then repeated in Italian to underscore the concept or the metaphor; in other parts of the country the reverse is done, as if to reassure their own credibility: the informal and the official, the workaday and the formal thus separate; often they oppose each other, integrating as signs of a frontier or relations.
Even poetry (particularly in Italy) is expressed in the local tongue, often with superb results: from the Roman Belli to the Milanese Porta, the Neapolitans Di Giacomo and De Filippo to the Venetian Noventa and Valentinis from Fruili. The writings of Pavese from Piedmont, Gadda from Milan, Bassani from Ferrara, Bigiaretti from the Marche, the Calabrian Alvaro and the Sicilian Tomasi di Lampedusa all feature the terminology, the syntax, the legends and the *humour* of their local context. On a more modest note, during the 1940s the author studied with his school companions with commentary in the Turin dialect – without offence to our classmate from Apulia, descendant of Abignente who, together with Fieramosca, was one of the famous twelve of the Disfida of Barletta – by the way, a battle celebrated by Massimo D'Azeglio from Piedmont.

Frontiers and cultures

The many 'internal' frontiers – often more than the differences in climate, local resources, periods of foreign occupation – have had a profound influence, as seen in the great diversity of customs, proverbs, foods, the way people dress, behave or greet each other, the makeup of the family, the status of women, the ways of love, the importance attributed to money, as well as the respect for work and the names of streets (recalling the trades formerly performed there). It is even said – and often quite rightly, if we consider the events between the 3rd and the 10th centuries, but also between 1944 and 1946 – that in Italy, occupiers have had a far greater impact on customs than the occupied. It is true that today the phase of defence of 'traditions' has begun – in Italy (as elsewhere), for example, museums dedicated to peasant lifestyles are sprouting – and this is something to be welcomed, even though the differences, the salt and honey of daily life, have been lost forever. Unfortunately however, there are also those who actually invent traditions and legends, perhaps to denigrate useful contrasts into small-minded divisions.

As will be nonetheless shown later on, the common knowledge and traditions of peoples inhabiting in proximity (beyond the chauvinism preferred or imposed for centuries), the marriages, friendships, wars fought together, even the mutual envy and the determination to overcome it, all favoured the recognition of our neighbour as a subject of possible – as some say today – 'collaborative competition', leading to exceptional achievements as soon as the opportunity and the thrust were given to overcome a condition of under-compensated fatigue.

For this reason as well, Italy, in the transition from a prevalently agricultural to a prevalently industrial society (even when there were no driving, advanced skills, such as a weapons or sea arsenal, in place since time immemorial, or special resources that led to their development, such as a mine, a crop type or an energy source), has become a prime example of a structure divided into *districts*, an inextricable intertwining of culture and work. Whether the enterprise was started up by a member of the nobility or a landowner, old or new bourgeoisie, a tenant farmer or his son (three well-studied modes), one labourer out of thirty or fifty eventually created a new enterprise. The phenomenon extended vertically to integrate, or horizontally to compete.

Farming implements typical of the Marche Region, early 20th century.

The Englishman Alfred Marshall (1842-1924) could hardly have had Italy specifically in mind when in 1890 he introduced the concept and the term, but it was here that the practice of districts experienced its most successful and convincing moment. Italian economists and experts on Italy distinguish them from *system-areas*, though acknowledge the excellent performance in terms of product, process and flexibility of both. Districts and systems are possibly the best proof of the importance of the so-called 'excellence curve'. Overall there are just under a hundred: from Brianza to Lumezzane and Manzano; Prato to Carpi and Sassuolo, to the Bologna area at the cutting edge of the *packaging* industry; from the Marche Region, Castelfidardo and Tolentino, to the Apulia Region[1]. Despite the repeated warnings by the financial press – which, comparing our tiny home-grown enterprise with the ones described by academic *management* texts, criticise its cultural shortcomings or the prevailing lack of larger industrial complexes – districts and systems alternate (just like IBM or FIAT, or the State chemical industry) periods of rapid and slow growth. And yet even before the pessimists have finished doling out their dire predictions, the situation has already changed or even greatly improved. We certainly have no intention of justifying shortcomings or negligence of specialised studies; but the fact is that, despite their poor *marketing* and chronic under-capitalisation, systems and districts help develop (if not at the right time, at least in time) innovations in products and processes, perhaps for short-lived initiatives but based on a solid idea. In the meantime, the capital necessary for development is mainly raised by *routine* business, given that the practice is to reinvest any excess without ever 'biting off more than you can chew'. The practice is all the more followed the smaller the departure capital; the problems were to arise later, with the attempt to put unearned revenue with income, or the need to reassure banks to obtain trading, then development, capital. It is a well-known fact that the *Peel Act* (1846) in England was not universally accepted; in Italy people say that, all too often, banks open their umbrellas when it's fine, then ask how much credit you have when you ask for a little. No matter how big or small the enterprise was, the 'new' was often discussed (and still is) at the inn or at home, in the sitting room, between master and worker, supplier and purchaser of a machine, a material, a semi-processed product – quite possibly culminating in a gentleman's agreement, not unlike what takes place in the squares of Mirandola or Asti on market day, when merchants spit on their hand before shaking for a deal. The mouth-to-ear exchange was (and still is) one of the main means of cultural communication, though it is today seriously threatened by new information technologies, by reducing the importance of geographic proximity, potentially informing anyone of anything – thus turning the technological mediation between distant and often faceless persons into a handsomely-paid profession. But nowadays little attention is paid to all this, as if it were not one of the factors that we risk considering as one that will change the world.

Lorenzo Lotto, *Virgin and Child with Saints Dominic, Gregory, and Urban*, 1506-1508. Recanati Civic Museum.

Men, women and families

Inside districts and systems there are men and families of different extraction, background, wealth. Families that are so well studied by Sergio Anselmi that even in this book he outlines the characters, not all evolutive, for the Marche Region as well. If of peasant extraction, the families; some from the North, closely related with father, mother and children from the modest, poor farmhouse; and in Emilia, Romagna and further down, made up of a number of extended family groups around a single hearth. Vast, multifaceted, complex, like a *clan*. As seen in the rolling pin for making pasta – short in the case of small families, or long, for use by two or more. Where the oldest – but not elderly mother – is addressed with terms of respect ("*resdora*" or "*vergara*"); she is the regent. Yarns and nursery rhymes of winter stables – an age-old chant recited as an ingenuous analytical taxonomy – "Il pane non è minestra, le pietre non sono mattoni, le nuore non sono suoceri, i preti non sono frati " – where wood was carved, wicker was wound, hemp was woven, and where even the men worked needle and thread, for just a bit more money outside the share crop contract, to pay the merchant's tribute. Crafts already described with erudite care by Tomaso Graziani da Bagnacavallo di Ravenna (1549-1589) in his remarkable register on "*Tutte le professioni del mondo*" ("*All the professions of the world*"), recently rediscovered after more than three centuries of oblivion[2]. These are today no more than memories, but help explain the bond, the continuity between the work, culture and life that characterises productive districts and systems, constituting its 'entry barrier'. *The future has an ancient heart*, a novel by Carlo Levi, Italian writer.

Certainly, the dissimilarity recognised between the industrial vivacity of the plains and that of the hillside[3] is not foreign to the need to compensate the scanty

Territory of the Augustinian monastic province of San Nicolò da Tolentino (corresponds to the modern-day Marche Region) (1639).

lifestyle offered by meagre crops of steep lands, which only today fully reveal (to the Wise of the European Union as well) their other side: that of quality, to finally reap their (mainly) economic benefits. In any case, more detailed studies[4] show how, in Italy, the industrialisation process has been characterised, at least until the second post-war period, by a long "dwelling on the threshold of transformation" – a slow yet constant flow that, while drawing vital energy from the long artisan-based (and financial) history of different parts of the country, had an overall effect of hampering accumulation and therefore development.

A frontier Region

In Italy, the term "marca" once denoted a frontier, from the German *mark*, used for border countries during feudal times. Pippin and Charlemagne in the second half of the 8th century donated several "*marks*" to the Church, which found there the frontline of its unrelenting confrontation with the Empire; usage of the term spread thereafter, until 1815 when the current plural form ("Marche") became official. A land of contrasts and tension, therefore – so varied in peoples, traditions and landscapes: from Romagna and San Marino in the north; from Tuscany, briefly contacted by the Apennines in the north-west; from Umbria in the south-west; from the short length of border lands with the Latium Region; and from the Abruzzo in the south. Hugged to one side by the Apennines, to the other the Adriatic – that large basin of the Mediterranean relegated to a 'minor' role when Venice, formerly vying with the proud city of Ancona, lost most of its importance as a trading port – the Marche is a Region characterised by paradox. Despite its enormous importance in Italy, the Region is elsewhere rarely mentioned, at least as such (despite the more than 100,000 replies on any Internet search engine). Many foreigners say "I'm going to Tuscany" or to Umbria, long preferring to travel to Rome via Florence. Few however mention the Marche, even though during summer, the contradas preferred by Cesare Borgia (otherwise known as "il Valentino"); the Urbino of Raphael the painter and Bramante the architect, the magnificent works of Laurana and Francesco di Giorgio Martini (commissioned by the duke Federico da Montefeltro) and the prison of Giuseppe Balsamo (known as Cagliostro), dank inside, delightful outside; the Pesaro of Gioacchino Rossini; the Recanati of Giacomo Leopardi, a

Lorenzo Lotto, *Miracle of St. Peter the Martyr*, 1506-1508.

Recanati, Piazza Giacomo Leopardi.
1998.

Historic centre of Urbino.

coastal, perched town – in short, the beaches of the Marche Region – teem with German, Dutch and French tourists. Alas, they visit the inland sites much less, where they could admire the masterpieces of Crivelli at Ascoli, or of Lotto in the tiny Monte San Giusto (Belli habitually spent his vacations in nearby Morrovalle); or the long, noble square of the Roman *Septempeda* destroyed by Totila (today San Severino, home of famed anatomist Bartolomeo Eustachi); Osimo, the Roman "*Auximum*", twice held for fifty years by the Longobards; the Macerata of Sferisterio and the Museo della carrozza, the monastic church of Tolentino, the recently rediscovered Roman theatre of *Urbs Salvia*, razed by Alaric; the uncertain heights of the Metauro, fatal to Hasdrubal; or the Montegranaro shoe factory in Ascoli where, after the Romanesque and Gothic periods, Annibal Caro, 16th century translator of Virgil, was prior; or Fabriano, with its famous cards. Not to mention the 126 museums and libraries[5], or the churches, featuring the severe Romanesque of the fishermen of Santa Maria a Portonovo or the mysterious ambulatory apse of Santa Maria a piè di Chienti, or San Claudio with its two towers, or the Cistercian abbey of Chiaravalle di Fiastra, or the marvels of Renaissance masters Sangallo and Sansovino, through to Vanvitelli in the Loreto of the devotions and *souvenirs*. But to continue would be to outdo the Baedeker guide. Armando Ginesi, in this very book, depicts the setting and landscape, restoring to the signs the meaning given by the syntax of their context, almost translating – like his beloved fellow-countryman and author of the "*Infinito*" – the 'site' into 'situation'.

In this Region, the original meaning of frontier – which always in a certain way implies dependence on another party – has strengthened that of the family, large or small, as well as that of technical collaboration, as seen in the unusual hydraulic system that as early as the 16th century was used to operate a mill, an oil press, a hammer for rags and paper, for metal working and dyeing[6]. There, the relatively slow development of activities such as forestry (which for water had supplied Rome since the days of the Empire through to modern times), farming or mining (in particular sulphur and stone) led to the progressive emergence of highly-refined crafting skills; in its small territorial dimensions, polycentricity (otherwise known as "pluralism") generates a high degree of social control favouring the longevity of useful traditions based on individual discipline[7]. The few, minor opportunities for 'nearby' exportation, i.e. just over the Adriatic, in the meantime procured not only a heightened interest in local and national resources and markets, but also more attention to 'distant' opportunities; initially in the supply of materials different from the local ones (in type, quality and price) or those less and less available – such as wood, in depleted supply due to extensive deforestation – to offer the local craftsmen; then for the export of manufactured goods.

In the famous letter (1825) that accompanied the inquiry by the Apostolic delegation on the products of the earth, as well as in the sheet annexed on manufacturing industries by gonfalonier Monaldo Leopardi – whom historiography strives to exempt from the severe judgement of his unhappy son – mention is merely made of "a small factory producing crockery and ordinary terracotta objects", "blankets of sheets or cloth (made by the lower people)", the "spinning and weaving products [...] that were marketed in Rome" and the "many workshops and small factories of hair combs and other crafted horn products"[8]. According to Monaldo, sixteen "artisans" – and overall 36 workers or hands paid by the day – modelled ox and sometimes buffalo horn, for a total of (at least) 15,000 horns

Stages in the crafting of horn cutlery at Fratelli Guzzini plant, 1950s.

from the communes of the province, Rome or Dalmatia; 33,672 combs of different size and design were declared, though there were actually no less than 90,000. Nor did it escape the attention of the person making the report that the horn rejects and pieces were allocated to the production of umbrella tips or cane knobs, if not for horse combs, made with lower quality material. In just a century, the Capodaglios of the Castelnuovo district in Recanati employed one hundred workers, while outsourcing part of their activities. In the early 20th century, a cooperative of comb makers was formed, and which under the supervision of Pietro Principi imported as many as 14,000 horns, selling some 150,000 crafted products; exports were also growing, as the Marseilles exhibition bears witness.
Immediately following the First World War, conditions developed at such a rate in certain industrial sectors that – rousing the people of the Region from the "patriarchal contentability" denounced by Angelo Celli in 1904, in his celebrated speech to the House of Deputies – they fostered the birth, organisation and growth of manufacturing activities soon to achieve significant results.
These activities grew in importance during in the last few years before WWII, as they were characterised by a high intensity of micro-collective work, product and process innovations, as well as a typological diversity that was all but unique. These developments were particularly surprising if we consider that, locally, macro-collective infrastructures and public facilities had never been particularly effective. Take for example the fact that, until the late 1960s, electrified railway lines ended north of Rimini (in Romagna) and south of Pescara (Abruzzo), creating a logistic void that the Region would pay for dearly. Indeed, the Cassa del Mezzogiorno – the public development fund in favour of economic development of southern Italy – was applicable up to San Benedetto del Tronto, in the Marche, just a few kilometres north of the Abruzzo border.
While the intensity of the innovative force (recalling the "creative destruction" of Schumpeter) of individual businesses and the appearance of *phenotypes* – variations of the same *genus* – may have seemed surprising, the number of *genotypes* on which the Region operated (and operates) is no less so. Sergio Anselmi's thesis on the socio-economic plurality of the Marche is thus confirmed in the technological diversity of the Region, as shown by the long lists of business sectors currently active: a range of productive activities that could even be associated with the careful yet aggressive strategy (not always successful in the long term) of farming opportunity (and low risk) identification adopted by landowners and tenant farmers. Here arises another paradox: a Region, whose very name recalls the idea of frontier, has today become not only a geographic *centre*, but almost a paradigm, of modern Italy. A country that, well after Pirenne's *Mohammed* and *Charlemagne*, once again emerges as a 'mark' of the Mediterranean; by the sea that unites and the mountains that separate, on the frontier of an ethnic, cultural – and therefore political and economic – future of the whole of Europe.
On the 20th July 1944, during the confused, testing period of the war front on the nearby Gothic line, twelve small businessmen (millers, pasta makers, engineers, builders, transporters, entertainers, cabinetmakers, brick makers and others) united to found the Associazione degli industriali della provincia di Macerata, the statute of which was approved the 27th December of the same year.
This industrial association was created with the primary objective of sub-Regional and Regional economic development. It was to be led by three important names in the industrial history of the Region: Scarfiotti (1959 and 1960, cement and chemical industry), Bontempi (1967 and 1969, musical instruments) and

Guzzini - Giuseppe, 1978, and Adolfo, 1987 – two of the grandchildren of founder Enrico Guzzini (1863-1948). Mention should also be made of Raimondo Guzzini, president of the Associazione Piccola e Media Impresa in 1966-67.
Enrico Guzzini – son of a tenant farmer of the Leopardi family in Recanati[9] and himself a former farmer – passed away, leaving three children, in the early years of the association, after founding the horn crafting company in 1911. Working also for other parties, he had extended the local product portfolio, mainly through the machining of snuff boxes, introducing new, more advanced French work tools, which he also offered to the Capodaglio catalogue.
Followed by the relative associations of Pesaro, Ancona then later of Ascoli Piceno, the creation of the association in Macerata was immediately announced by the Manufacturers' Association of Rome (January 1945). The 249 industries employing 4,020 workers registered in 1945 had risen by 1949 to 484 industries and 6,338 workers, more than half of those surveyed in the provinces (in five years, an increase of 94% in the companies registered, 58% in workers). Nonetheless, in 1951 farming accounted for no less than 60% of the working population of the Marche, dropping to 45% only ten years later, largely due to emigration from the Region. Most industries were still closely tied to agricultural products, constituting an immediate 'secondary' sector for processing - mills, oil presses, cellars – as against final consumption. Only in the early seventies did the number of workers in industry actually exceed those in agriculture; nationally, industry had overtaken agriculture 15 years beforehand. The 1961 census revealed that the per capita income was the lowest compared to the Italian average since the war, while product volume per active work unit was the smallest in the country.

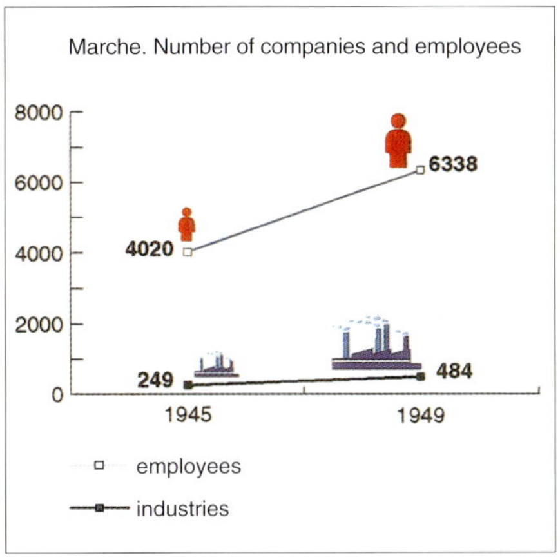

Quarters of nobility

At least eighty years before Enrico Guzzini, a group of entrepreneurs[10] – nobles such as Emilio Ripanti (paper), Aurelio Balleani of Jesi, Andrea Buffoni of Fossombrone (silk), Luigi Merli, then grandson Francesco Luigi, pioneers of electrification of Ascoli; middle-class or children of tenant farmers, such as spinning mill owner Pasquale Mancini of Jesi, Giovambattista Marotti and Edoardo Almagià, builders of Ancona, liqueur producer Silvio Meletti of Ascoli – had paved the way towards organised business activities.
Excluding Marotti, who oversaw important railway and port construction works, and the above-mentioned Capodaglio, business activities were prevalently concentrated near the land and on age-old skills, with mixed results, though many successful. The exception was Paolo Soprani (1844-1918) of Castelfidardo, who (since the very year of Enrico's birth) at the age of 19 started up a company in a 'different' field of goods, characterised by its unusual degree of structural complexity: the piano *accordion*, inspired by the Central European models. (Something similar to the adventure of the Novarese Giuseppe Ravizza, who ten years earlier had designed the "*cembalo scrivano*", a prototype of the modern typewriter. However, while Soprani had immediately set up an enterprise for his intended activity, Ravizza stopped at the design stage, never actually seeing his ideas revised and carried through to production by Camillo Olivetti in Ivrea at the end of the century.)
Researchers of Italian economic development would do well to extend their attention beyond the "long, narrow half-moon, concerning the hills, plateaus, the mouth of the alpine valleys to the lands of Piedmont"[11].

Interior of piano accordion box, with reeds and closing mechanisms. International Piano Accordion Museum, Castelfidardo (Ancona).

Enrico's generation was rather that of Sisinio Fagioli of Osimo (building and furnaces), the earth-moving machines of the bold Giuseppe Guerri and the mechanic Adeodato Pieralisi of Jesi, the tanners Lorenzo Feroci of Monte San Giusto and Nazareno Gabrielli of Tolentino, the furniture of Renato Fastigi of Pesaro, the paper mills of Giovan Battista Miliani in Fabriano.

As can be seen, the range of business activities was expanding, gradually moving away from those directly associated with the land and farming as they sought local or national markets. New attention had to be paid however to international competition, in particular following the denunciation of the Italian-French trade treaty (1886), the beginning of the so-called 'tariffs war' that would only end a dozen years later. In 1890, the Zanardelli code had sanctioned the freedom to strike and the freedom of non-strikers to work. The following year saw the publication of the encyclical *Rerum novarum*, immediately enacted by Italian prime minister Giolitti. In 1886, when Enrico Guzzini's generation had still not reached thirty, a new 54-year macrocycle[12] began and that would only conclude with the start of the Second World War (1939-45), today recognised as the second half of the First. The transition from one cycle to another was marked in Italy by the crisis of 1880-90, when the stagnation of national income led to a dramatic drop in per capita income; "after having paid the previous decades in cash, the countryside pays again on credit"[13].

Notes

1. In 1992, the *Edizioni* of the *Sole 24 Ore* collected a series of monographs published by the newspaper: *Gioielli, Bambole, Coltelli,* edited by M. Moussanet and L. Paolazzi. Despite the confusion between districts and systems, the local kaleidoscope of Italian industry emerges in all its colours.

2. *La piazza universale di tutte le professioni del mondo* (first ed. 1585, last ed. 1675), edited by Paolo Cherchi and Beatrice Collina, 2 vols., Turin, 1996.

3. See Franco Bonelli, "Il Capitalismo italiano. Linee generali di interpretazione", in *Storia d'Italia*, Einaudi, 1978, and Alain Dewerpe, *L'industrie aux champs. Essais sur la proto-industrialisation en Italie du Nord (1800-1880)*, 1985.

4. Cf. Dewerpe, *op. cit.*; Luciano Cafagna, *L'industrializzazione italiana: la formazione di una "base industriale" fra il 1896 e il 1914,* 1961; and *Proto-industria o transizione in bilico?,* 1983.

5. The Marche Region hosts an enormous number of museums: 39 centres in the province of Macerata, 32 in the province of Pesaro-Urbino, 29 in the province of Ancona, 26 in the province of Ascoli.

6. *Appunti storici sulla cartiera di Ascoli Piceno,* Circolo ricreativo ENEL, 1978 (see also Michele Dean, in *Le Marche*, edited by Sergio Anselmi, Turin, 1987).

7. See Sergio Anselmi, "Marche", in *Stato dell'Italia*, Milan, 1994.

8. See Franco Foschi, "La manifattura dei pettini di corno a Recanati e sua evoluzione", in *Proposte e Ricerche*, Urbino, 1983, fasc. 10, p. 109 ff.

9. The Recanati of the poet of the Infinito – like all towns of the Marche Region featuring a castle – has a long history dating from the time of Imperial Rome and the Greek-Gothic war. An independent commune as of the 11th century, the town was awarded the right to a port (Porto Recanati) by Frederick II; siding with Gregory IX in the famous conflict with the Emperor in the early 13th century, it was granted the title of city with seat of the bishopric. Gregory XI died there in 1417 after renouncing the papacy. Four years later, Martin V instituted a fair that was to long compete with that held in Ancona. Reunited with Loreto in the late 16th century, the political importance of the town began to decline.

10. See Franco Amatori, "Per un dizionario biografico degli imprenditori marchigiani", in *Le Marche*, edited by S. Anselmi, cit.

11. See Alain Dewerpe, "Verso l'Italia industriale", in *Storia dell'economia italiana*, Turin, 1991; vol. III, p. 26.

12. Reference is made to the 50-year "long waves" by Russian Nikolaj Kondratiev (1917), more recently revisited and corrected in 54 years by Italian Cesare Marchetti, IIASA, International Institute for Applied Systems Analysis (Laxenburg, Vienna). Cf. "Society as a Learning System", paper IIASA, 1978.

13. See Alain Dewerpe, "Verso l'Italia industriale", in *Storia dell'economia italiana*; Turin, 1991, vol. III, p. 34.

From the countryside to industry
Sergio Anselmi

Map of the Marche Region, 1831.

When the extent of the changes witnessed in the world of the post-war era became evident, particularly in the latter part of the 1970s, some began to wonder which factors (besides the regular ones) had so radically transformed the Region that goes under the name of the Marche, making it the first among *second comers* in Italy.
A rhetorical question, because the situation was clear: all you needed to do was open the windows and take a look outside. Preclusions and ideological conditioning (exploitation and self-exploitation, illegal labour, fiscal evasion, dishonest traders, small-mindedness, etc.) however remained, leading to on-going debates regarding the ugliness of the Marche model, its low profile, its intrinsic fragility[1].
The reality did not correspond to the theory, so it became necessary to change the representation, pretending to not see it as it actually was.
Fortunately things proceed according to their order, which does not always correspond to the one conceived by many academic and trade union schools; and so the Marche economy could make its great leap forward without serious upheavals[2], indeed in conditions of social peace, with however the inevitable costs that all great transformations[3] entail. Academics from different cultural backgrounds (economists, sociologists, historiographers such as Giorgio Fuà[4], Massimo Paci[5] and Sergio Anselmi[6], with their collaborators distributed throughout various universities in the Region, in Emilia-Romagna, Tuscany, Umbria, Abruzzo), though basing their ideas on different suppositions, came to the same conclusions, thus confirming the reflections of Francesco Caletti (1914-23)[7] and Bruno Ciaffi (1953)[8]. They identified the driving force of economic development in the tenant farming families of the period 1955-75 and the small business enterprises located in the 246 communes of the Region and their administrative emanations. All achieved with endogenous capital and labour, without large-scale state intervention (with the exception of the Aprutino-Ascolana strip) and with modest infrastructures; communications systems were indeed still those typical of farming land, with the inhabitants spread far and wide over mountains, hillsides, valleys, coastlines. An overall population of around 1,400,000 distributed over a Regional surface area of 9,694 km^2 – equal to 3.1% of Italian territory – and a density of 145 inhabitants per square kilometre.
The first post-war census of 1951 showed that two thirds of the people lived in disparate homesteads and in rural woods, with a rate of 60.2% involved in farming, one of the highest percentages in Italy. No large cities. Even today, the Regional capital Ancona has fewer than 100,000 inhabitants, while the other major towns have populations between 30,000 and 90,000: Pesaro, Fano, Senigallia, Jesi, Falconara, Civitanova, Macerata, Ascoli Piceno, Fermo, San Benedetto del Tronto. It is, however, also true that the Marche is home to no fewer than 18 centres of ancient and solid housing and economic centres (Castelfidardo, Chiaravalle, Cingoli, Corridonia, Fabriano, Fossombrone, Grottammare, Loreto, Montegranaro, Osimo, Porto Sant'Elpidio, Porto San Giorgio, Potenza Picena, Recanati, Sanseverino, Sant'Elpidio, Tolentino, Urbino), contributing thus to the creation of a second urban network with an even finer fabric than the first, though more extensive than those of centres that could be defined "*third best*".
A territory that is today thoroughly interconnected, even if each habitation is affected by the limits of the railway system and roadways – certainly more modern, but still conditioned by the ancient routes traced by road communications, with bottlenecks around the port of Ancona, not to mention inadequate air con-

nections to and from Falconara, the only Regional aerodrome.
The explanation: the tenant farmers and artisans, notoriously thrifty even when financial resources are readily available. The general climate of reconstruction and the market also create the conditions for development: the peasant population, leaving behind 106,000 country homes and the fields on which they had lived, finally felt ready to head to the suburbs, using the knowledge accumulated over centuries of ingenuity applied to a wide range of activities, essential to the self-sufficiency of the family, in which money was to enter and not to leave, work was distributed and performed according to the principle of "how much you consume depends on how much you produce", adults ("bocche da sale") provided for their own children ("bocche da latte") until they themselves (very early) could produce; respect and deference to the elderly remained the rule. In the hard school of domestic life, everyone had a role to play, whether in the field or the household – a place of residence and labour, where besides the care for livestock and "dead" stock (tools), not to mention attentive management of food supplies, fabric, baskets, carved wood, straw products, clogs and even slippers were produced for the market. As well as everything that was needed to keep the household and farm running: from masonry work and fittings, the wooden parts of work tools (the metal parts were usually reused), fencing, shelters, sewage, through to the maintenance of access roads to the home and fields.
The family worked on other premises on a standard (or "basic") tenancy contract, according to the principle of *fifty-fifty* sharing of expenses and revenues – a system quite different from the simpler one of sharecropping, which proved unsuited to the modern situation in the Region[9]: in the latter case, the farming family was made up of the elderly parents (the "vergaro" and his wife the "vergara", also known respectively as the "capoccio" and the "capoccia"), the "young" couple (husband and wife) and their children. In actual fact the size of the family depended on the dimensions of the land to be worked over low and medium hillsides (in the Marche, plain lands and mountains are almost inexistent), on an economic principle "one man per hectare", or in more technical terms, "one ox per hectare".
Given that the average size of the 99,105 farms (official register 1929) was around 6.33 hectares, then rising to approximately 103,000 hectares over the next twenty years (due to the break-up of farms larger than 20.01 hectares[10]), before the diaspora of the sixties and seventies, tenant families were presumably made up of 6-7 members, reaching 20-30[11] and more. The larger families often took on a patriarchal character: besides the elders, there were several "young" couples (brothers with their respective wives), unmarried daughters and sons, sometimes a farm hand.
These trends are confirmed by the official figures regarding the average farming and town family in the Marche Region: 1931 = 5; 1936 = 5.1; 1951 = 4.5; 1961 = 4; 1971 = 3.6; 1981 = 3.1; 1991 = 2.9[12]. The latest figures show an alignment with the Italian national average (in 1997 2.6).
Until the 1950s and 60s therefore, there were over 100,000 farms with just as many farmsteads, equal number of capacity, working families or production nuclei with work distribution, integration and internal mobility regulated by consolidated procedures skilfully applied by the firm hand of the "vergaro" (the master, who wielded the symbolic "verga" or staff of command) on what were actually and by rights 100,000 enterprises, in which capital (the land) was rented out to family work groups led by a "foreman"; enterprises required to pro-

Marchegian farming family
assembled in farming house yard,
early 20th century.

duce revenue for the "farm" and profit to the "workers" by the end of the farming year.
The departure from the countryside of this mass of men and women, accustomed to a frugal lifestyle, consuming little more food than the energy expended during work, disciplined, hard-working and pragmatic, could not but have a remarkable, even visible, impact on the entire Region.
The suburbs began to sprawl around the valley towns, joining up the coastal localities into long villages, transforming the farming scene, increasingly in the hands of "terzisti"[13] who, to make their activities economically viable, were forced to eliminate trees, hedges and ditches, thus making the terrain suitable for agricultural equipment.
The first impact was objectively strong: entire lands and towns emptied, countryside virtually deserted, industrial sites strewn all over the territory, a coastline covered by hotels and pizza restaurants similar to the model seen in Romagna, makeshift (often incomplete) housing according to the shapeless urban-rural design of house/vegetable patch/warehouse – a far cry from the sober elegance of the now-abandoned country homesteads.
However these modest parallelepipeds, in which everything was sacrificed in the name of pragmatism, witnessed, in the shade of the still-influential patriarchal contentability, the beginnings of professional diversification: manual trades, retail, apprenticeships, services, handcrafts, piecemeal housekeeping.
Having left behind the old farmhouse and with it the parish community (the home of masses, baptisms, weddings and funerals), the family began to explore the ways of the world, accepting the ideas of compulsory education, that women could refuse to clean the stables, of the possibility of competing with more enterprising citizens, who worked at higher costs than those that these former farmers could offer.
The precariousness of these new town, suburban and coastal settings was initially an eyesore, with factory sheds strewn haphazardly over the countryside (often thanks to the support of parish authorities and work foremen), the loss of attractive sights, gardens, Liberty-style villas, the underhanded assault on castle towers and walls, gross attempts at architectural camouflage. Towns had no planning schemes for private housing, business activities, roadways, public services. Only in a few cases were attempts to contain this hyperactivity successful, which completely devastated certain hilltop towns set amidst secular trees. Or with the pretext of defending the "historic township" or the "ancient convent", the interior was sacrificed, thus permanently degrading the entire structure, which thus lost its intrinsic original beauty.
It is possible that these events are not unique to the Region; nonetheless, for the gracious Marche of tenant farming, of the "bel paesaggio" noted since the 16th century[14], the transformation, a veritable headlong dash towards change, had a violent impact, though the most appalling aspects were somewhat mitigated not so much by law as by cultural "sedimentation": that of landowners, tenant farmers and artisans, concerned by pressing current needs, but also (due to the influence of family and community values) of the future of their children and grandchildren.
In the early eighties, as in other more consolidated sectors of the economy, a turnaround began, brought on by generational transition, the emergence of new tastes, a new awareness of the importance of image (in monetary terms) that houses, factories, public buildings, avenues, parks, tree-lined suburban streets

Preparation of feed sacks in a Marchegian industrial plant, early 20th century.

took on in the affluent society knocking at the door. The Marche model – typified by the small and medium-sized businesses distributed throughout the Region, the neatly-arranged towns and villages with their restored historic architecture, the rural landscape dotted with farm houses converted into country residences for people tired of the city, the less tilled and more tree-covered fields (of the 915,000 hectares of farming land available – and almost completely used in the past – a mere 550,000 are now used, without any reduction in overall productivity) – has come to the attention of economists and architects alike, also due to the average size of family savings recorded, estimated at almost ITL 5 million/year, against ITL 2.9 in Umbria and ITL 3.2 in Tuscany[15], the other Italian Regions in which tenancy farming was just as intense and widespread.
Some people, referring back to the old expression coined by Ghino Valenti (the *aurea mediocritas* of the Region[16]), then adopted by Francesco Caletti ("a widespread agricultural community with average, well-balanced features"[17]), have spoken of the "modernization reduced by half"[18] (but "halved" compared to what modernity? And what are the measuring criteria suited to the task?), because in the Marche "the asynchrony [....] between rapid economic development and slow, risky cultural transformation"[19] is marked by the prevalence of "familism and localism"[20], which are then counted as factors of growth.
The Marche Region today is visibly different from that of yesterday, though it is still recognisable by its quality of life, as well as its landscape. Could and can things be done better? The overall situation (people in feverish movement and attached to the territory) is no longer as alarming as during the "leap forward", at the cost of the environment as a public asset. The basic structures have stood up well. By the same token we cannot ignore the principle that the current environmental set-up is probably not forever. It will always pose new challenges, taken up by sensible people with the awareness that no final solutions actually exist; in other words, a state of stable balance cannot last indefinitely.
The people of the Marche, neither Levantines nor Prussians (but maybe a bit of both), are skilled in combining efficiency with the quality of life; and it is possibly for this reason that many businesses aim at growing even more, by refining their productive organisation, to conquer greater market shares, without betraying their own origins.

Notes

1. For an initial insight, see the discussion between R. Schiattarella, "La piccola impresa e le Marche negli anni Settanta", in *Proposte e ricerche*, 18, 1987 pp. 113-124, and A. Niccoli, "Lo sviluppo dell'industria marchigiana: quale o quali interpretazioni?", in *Proposte e ricerche*, 19, 1987, pp. 194-198. In 1984, Niccoli had tackled the issue of the origins and development of growth in the article "Alle origini dello sviluppo economico marchigiano", in *Economia Marche*, 1, 1984, pp. 3-17. As regards the "remote" and "present" past of industrialisation in the Marche Region, see also E. Sori, "Dalla manifattura all'industria" (1861-1940), and C. Zacchia, "Il quadro economico Regionale dal dopoguerra ad oggi" (1987), in S. Anselmi, *Marche*, in *Storia d'Italia. Le Regioni*, Einaudi, Turin, 1987, pp. 301-392 and 393-423 respectively.

2. G. Fuà, C. Zacchia (edited by), *Industrializzazione senza fratture*, Bologna, 1983.

3. Karl Polanyi, *La grande trasformazione*, it. ed., Turin, 1974. In the self-regulating market, the "great transformation" turns work, land and money into "fictitious goods" (pp. 93-94).

4. Giorgio Fuà, founder of the faculty of Economics of Ancona and chairman of the Istituto Adriatico Olivetti (Studies for economic and corporate management), full professor of Economic Policy, was the first to theorise the "Marche model" and the inventor of the fortunate acronym NEC, indicating the development of the North-East-Centre. He was also responsible for the creation of the six-monthly magazine *Economia Marche*, regularly published since 1976.

5. Massimo Paci, professor of Economic Sociology at the faculty of Economics in Ancona, was one of the first to identify the particular nature of the impact on the Marche social fabric in the wake of the post-war transformation (1945 to 1975 approx.) of the economy, in particular in micro-industry based on private enterprise, tenancy farming and home-based work activities: the family, women's rights, children, the elderly, "owner-users", the *welfare state*, etc.

Paci was also the first editor of the magazine *Stato e mercato*.

6. Sergio Anselmi, full professor of Economic History at the faculty of Economics in Ancona, studied the history of the Marche countryside, identifying classic tenancy farming as the point of departure of economic and cultural changes in the many centres of services and artisan workshops in the Region. In 1978 Anselmi established the six-monthly magazine *Proposte e ricerche, economia e società nella storia dell'Italia centrale*.

7. F. Caletti, *Il carattere rurale nell'economia e nello spirito delle Marche*, Piacenza, 1923, which refers to an article published in the *Giornale degli economisti* in 1914, entitled "Le Marche e i moti recenti [della 'settimana rossa']".

8. B. Ciaffi, *Il volto agricolo delle Marche*, Bologna, 1953.

9. As regards the development and refinements to "classic" tenancy farming (i.e. with stable residency of the family on the farm) in the Marche Region, 14th-20th centuries, see S. Anselmi, "Padroni e contadini", in *Marche*, cit., pp. 243-297; Id., "Mezzadri e mezzadrie nell'Italia centrale", in P. Bevilacqua (edited by), *Storia dell'agricoltura italiana*, 3 vols., II, Venice, 1990, pp. 201-259; Id., "Mezzadria e mezzadri", Senigallia, 1993.

10. B. Ciaffi, *op. cit.*, pp. 111-112 and 173.

11. *Ibid.*, pp. 179 ff., *L'Inchiesta agraria Jacini*, Rome, 1883, vol. XI, t. II, chap. XXI and appendix.

12. Istat, ad annos. In tenant families, the average rose to 7.5 (1921) – 7.6 (1931); see A. Cortese, "Modificazioni della famiglia marchigiana", 1861-1981, in *Proposte e ricerche*, 16, 1986, pp. 82-90.

13. "Terzisti" or "terziaristi" are those who on the behalf of third parties operate on land abandoned by tenant farmers and not worked by farm labourers. They usually exploit the lands rented from owners, making intense use of machinery and chemical fertilisers: tilling, sowing, harvesting, marketing.

14. Montaigne to Venetian ambassadors, struck by the beauty of the orderly countryside and the professional pride of the farmers. On the "travellers", see N. Cecini, *Le Marche: una metafora per i viaggiatori*.

15. S. Anselmi, "Dall'agricoltura all'industria", in Id. (edited by), *L'industria nella provincia di Pesaro-Urbino*, Assindustria, 1995, pp. 11-30, separately on p. 22. On this and other aspects of the great transformation (apiculture-industry) of the provinces in the Marche Region, see the recent writings of the same author (who apologises for the self-quoting): "Dall'agricoltura all'industria nel Maceratese", pp. 24-31, in *Cinquantesimo*, 1944-1994, Assindustria, 1994; "Dall'agricoltura all'industria diffusa", in A. Niccoli (edited by), *Fare industria nella Marca di Ancona*, Assindustria, 1995, pp. 33-54, and, S. Anselmi, "L'industria calzaturiera nella recente crescita delle Marche. Ipotesi storiografiche, problemi, linee di sviluppo", in Id. (edited by), *L'industria calzaturiera marchigiana*, Unione Industriale del Fermano, 1989, pp. 15-37.

16. For reference to Francesco Caletti, P. Sabbatucci Seonini, "L''aurea mediocritas': le Marche attraverso le statistiche, le inchieste e il dibattito politico economico", in Anselmi, *Marche*, cit., pp. 209-239.

17. G. Valenti, *L'economia rurale nelle Marche, Macerata*, 1988. Valenti is known for having written the monograph on the Marche Region for the Jacini farming inquiry.

18. C. Carboni, "Mentalità, lavoro e classi sociali", in Anselmi, *Marche*, cit., pp. 631-657: "An active Region, increasingly wealthy, tranquil. This is apparently the secret of the Marche, that in recent years has catalysed the attention of academics and economic policy makers. One nonetheless has the impression of modernisation decreased by half", p. 631.

19. *Ibid.*, p. 633-634.

20. *Ibid.*, p. 631-633. See also M. Paci (edited by), *Famiglia e mercato del lavoro in un'economia periferica*, Milan, 1980; and in particular the article by the same author "Riflessioni sui fattori sociali dello sviluppo della piccola impresa nelle Marche", in *Economia Marche*, 6, 1979, pp. 71-88.

The metayage culture

Sergio Anselmi

Modern Marchegian countryside.
Photo by Gianni Berengo Gardin.

The great transformation

"My name is Bonfiglio Andrenacci, from Sant'Elpidio, my family is made up of me, my brother, our two wives, the widow of [another] brother and 11 children, in all 16 members; we are of farming stock and we cultivate about 20 hectares as tenant farmers"[1]. This is how one of the numerous tenant farmers, paternally questioned to discover their daily behaviour and projects, described himself to the agents of the parliamentary inquiry on "the conditions of the farming class"[2]. He continues: "There is agreement in my home, and all members of the my family depend on me. I believe that this agreement and relative dependence are behind the quite good results achieved by cultivating the land"[3]; we eat "cornbread for about 11 months of the year, otherwise wheat bread, during harvest time; soup twice a day between April and the end of September, during the rest of the year once a day"[4], with annual family expenditure (actually spent on food eaten with bread, salt and taxes) of 1,870.40 lire for "maize, wheat, wine, pork, oil, salt, clothes, food eaten with bread, incentives and taxes"[5].
Another tenant farmer, Agostino Rogante, resident in the same period on a 26-hectare farm, near Fermo, declares: "Our family is made up of 24 members, four of which are married, a widower and a widow, two youngsters ready to take wife. Even though my family is large, we get by better than others [...and at the end of the year] I still have a small quantity left over, [... because] there's a lot of order in my family [...]. I look after everything, I try to provide for everyone's needs to the best of my abilities [... and I give] as much cornbread as they want during the two meals. In the morning, vegetable soup with a condiment of pork fat; in the evening, either herbs or flour soup [...] and I even give my family wine in winter, on Sundays, between Easter and September every day, though diluted with water [...]"[6].
Both the "vergari", like almost all of the people questioned by the inquiry commission[7], insist on a number of aspects of their day-to-day lives: the family, supported by a firm hand of the family chief, whose orders and those of the land owner must be obeyed[8]; unfortunately, feminine luxury, gambling, smoking and taverns are becoming widespread in the countryside[9]; loan money for purchasing a farm is too expensive, as much as 15% (although relatively few wish to become direct crop farmers[10]); damage is constantly being wreaked by the thieving antics of labourers or land-less peasants, the poor of a society of either citizens (artisans, owners, professionals, public servants) or tenant farmers[11]; the load of taxes and dues is excessive, especially those levied on livestock, flour or hearth[12].
Thus was the situation in the late 19th and early 20th centuries, when the rural family had to come to terms with two emergencies: a surge in population growth (in the Marche +236,490 between 1861 and 1911), and the total working of all available farming land. As one farmer from Fermo asked in 1883, "The population is growing every day, but where will we send them?"[13].
The reaction of a rural economy with consolidated characteristics (the home, food and some savings for tenant farms; revenue in the case of land owners; profit in the case of bailiffs and grain and livestock merchants, metal tool manufacturers) was inevitable: there is not enough work for everyone, so those not needed must leave the farms according to work requirements. There is only one path (expulsion from the family and the tenancy system), which eventually separates into the downgrading of farming labour (the so-called "casarolanti" forced to live by expedients, scorned by society)[14], or emigration. Between 1876 and

Bas-relief by Pericle Fazzini: model for *Il solco*, 1965.

1910, the Marche witnessed a growth in the annual rate of emigration of 0.32% (1876-80), 2.0% (1881-90), 4.77% (1891-1900), 20.57% (1901-10), 24.97% (1911-13), rising from 12th to 6th highest of all Italian Regions[15]. Emigration was a clear sign of a social malaise deriving from a rigid economy that, being consolidated and accepted by all those involved in it, becomes increasingly set in its own culture. Another sign of social malaise was the spread of delinquent crimes[16].

Though holding, the tenant farmer family was forced to come to terms with increasingly closer relations with the farm on which they lived and worked; large land owners, on the other hand, tended to break farms down into smaller properties to improve output, given that a family thinly spread out over great expanses was hard put to (and often unable to) attain a decent standard of living. In this way, a larger share of the harvest went to the land owner.

The opposition put up by many tenant farmers (particularly those working on the most productive estates) to the abolition of the status quo achieved is so strong that not even the social upheavals witnessed in the countryside had a significant effect in the Region[17], despite the conviction of most economists (especially those with socialist leanings) since the late 19th century that tenancy farming needed to be reformed rather than abolished[18] beyond certain options of principle.

The fact that farming families lived in harsh, rustic conditions is clear, but should also be viewed in relation to those borne by farmers in southern Italy, by workers in irrigation and mountain Regions, by labourers in the cities and seaside fishing communities, who suffered even lower living standards. Indeed cases are known of tenant families living on the very same farms for centuries (not to mention for a full millennium[19]), indicative of a relationship based on a traditional bond to the land and its owners.

The family of the tenant farmer

Tenancy relations were determined between the parties by means of the "scritta" or farming policy. The contracting parties included the lessor (property owner) and the tenant or the farmer, who also acted on behalf of the family, composed of a number of persons working to ensure the productivity of the farm. In the event of necessity, the farmer was required to employ a farm hand at his own expense. The owner provided not only the farm, but also a farmhouse suited to the requirements of the family, together with stables, warehouses, chicken coop, cellars, etc.. Working livestock constituted a *joint asset*[20]. In almost all cases, the bailiff or "minister" (or "agent") ostensibly mediated between the owner and the workers, even though considered by the latter more favourable to the owner; he was therefore feared, though respected for his technical skills.

The family chief (commonly known as the "capoccia" or "vergaro") directed the operations of all members of the family, from the fields to the stables. The wife of the family chief was in charge of domestic activities (cooking, clothing, dowries), the vegetable garden, poultry, eggs and, if present, sheep and cheeses. As a rule, families made up of a single branch were headed by the father; in the case of families with several families, "sometimes it's the grandfather, if he has the intellectual capacity; on the death of the family head, the eldest son (or other son if shown to be more competent in handling the assigned responsibilities) takes over. The successor is, however, always selected with the consensus of the family"[21].

Of the other family members, a special role is played by the most robust of the males (the "boforco"), who is in charge of the cattle, as well as of driving the plough, the cart or wagon (*birotum*), the barrow and the grass sled. The work implements, just like those for personal use, usually belong to the farmers themselves. If necessary, contingent exchanges of functions are possible, though always with total respect for the principle of formal responsibilities. The farmer family – except in the case of the "grandi fatiche", the large-scale operations, when all members are called to assist during hoeing, grain harvesting, threshing, grape picking – operates by minor tasks assigned by the family head or his wife, according to whether they are conducted in the field or around the house. As regards labour distribution, consumption is proportional to productivity (besides the minimum quantity required for sustenance). The various members of the family, ordered according to age and sex, were thus evaluated: elderly family head: work 0.50, consumption 1; adult male: 1:1; adult females: 0.66:0.75; female and male offspring: 0.35:0.75; young children: 0:0.50. Though quantification and distribution were not enforced rigorously, such calculations were common right up to WWII and even beyond[22].

If too many females, or invalid children, are born into the family, problems arise. Hopefully, the women marry early (and go to live with their husbands), even if the dowry is substantial. As early as the second half of the 19th century, the Jacini inquirers estimated the high cost of dowries in the Marche: from 1,629 lire for a young girl from a "large property", to 419.50 lire for a girl from a "small property", or other costs in between[23]. Even in the post-war period, the cost of women's clothing is high, because "women, especially young ones, love luxury, and during festivities dress well and with taste"[24], a sign of strength that they exert their condition as workers and labourers, in response to male hegemony, which sees "luxury" as an element of disorder.

Women participating in tenancy farming are just as efficient as men; indeed, during WWI when stronger men were in short supply, the production of food remained stable[25], though adding to their operative capacity the grace of their condition (jokes, flowers, ornaments), thus compensating in the farming household the male rusticity, while giving in to a certain coquetry. No moral upheaval, despite parental promiscuity in the farmhouse. In the rural premises[26], everything is clear and simple, though not essential: "Furniture was once made up of simple iron beds, or two trestles with tables and a straw mattress with corn

leaves, in drawers, sideboards, small tables, seats and other household goods; but always coarse; today (1953) however things have improved"[27]. But the tenant farmer, emerging with some money[28] and a few more ideas[29] since WWII, thinks of leaving. The *Lambretta* and middle school provide backing.
The only solution is to convert the exit into cash, make family divisions (old capital, new capital, seizure of crops and annual revenue), then go down the paths of the world. He will do anything but farming. Indeed the tenant farmers leaving the field and occupied by the most diverse activities do not include the new land owners. The refusal of the land is global, but not the cultural heritage in the diachronia of the families.
The war had interrupted the *continuum*, the next twenty years codified it.
It has been written that "people are born peasants like others are born nobles". Access to the land is not a matter of free choice; if anything, the land is abandoned for the city. Tenant farmers marry amongst themselves, subject to the control of a family chief who commands at home a degree of respect that is directly proportional to esteem enjoyed amidst other tenant farmers, bailiffs and property owners.
There are three measurements of esteem: the respect for the given word, the technical and agricultural skills, the family living standard, complemented by sobriety of language and impartiality of judgement. The good family chief is proud, and honours the memory of his ancestors, passed on to children and grandchildren through mottos, proverbs, expressions and example. He is the first to rise in the morning, the last to lay down at night. He dresses modestly in home-made clothing, but is never slovenly, knowing how to sit at the table when the owner invites him to breakfast at least once a year.
He assigns the "tasks" and no-one, not even his wife, thinks that he can be publicly contradicted. This does not mean that obedience is blind and absolute, but there must be good reasons to question any decision.
The boss, who (as is known) governs the home and courtyard animals, the vegetable garden and the loom (on which everyone must work during the cold seasons) is *naturaliter* parsimonious. He holds the keys to the pantry, looks after the education of the youngest children, reproaching their mothers if they do not behave seriously or argue with each other, ensures that the dowries of daughters grow gradually and justly, to be ready for the day of their weddings. The same attention is paid to the dowries of daughters-in-law. Even the dowries, exposed to relations and neighbours before the main wedding meal, confirms the reputation of the family in the territory.

Early mechanical farming equipment in the Marche.

Young men, long remaining between childhood and maturity, experience the thankless condition of the subordinate, which the patriarch sometimes mitigates by requesting advice, varying tasks if one is more unpleasant than another, distributing a half cigar or another glass of wine. Religion is formally practiced, more by women than by men, but is often characterised by the "materialistic, or even the egotistic with divine Providence". The parish priest, who sometimes visits the tenant farmers, is welcomed everywhere with respect.
Nonetheless "the manifest profession of faith and a heritage of superstitions are accompanied by a weakly-understood national sentiment [...], and yet the young willingly don the military uniform, but mainly to avoid the humiliation of a declared unsuitability to serve", because a saying was doing the rounds: "Anyone not good for the king is no good for the queen". Mother, the local parish priest and, as of the 1910s, the elementary school teacher, educate the young

Opposite page: farm in the Marche.
Photo by Gianni Berengo Gardin.

boy to persevere in "simple traditionalism, which represents one of the most salient characteristics of these people", who even then (1935) were paternalistically studied from an anthropologic viewpoint, assessed as good, hard-working communities. Young women performed occasional chores in the owner's home (e.g. the laundry), receiving in exchange a few personal benefits: old clothes, soap, lingerie. Elderly women wove hemp and cotton cloth for the landowner's wife, who sometimes requested firewood to be chopped and stacked, as well as the preparation of pig meat from the farm.

In the light of the range of activities performed by tenant farmers, attempts have been made to quantify the work taken by farms from agriculture; the attempts however have been largely in vain, because when good relations subsist between land owners and farmers, a tacit understanding is reached that no family would limit their activities, in the name of respect, as well as an unwillingness to reduce their 50% share of harvests. Why less refined property owners were unappreciative of the farming family's "supplementary activities" is understandable (particularly since they tended to strengthen the bargaining power of tenants), though it would be erroneous to interpret it as a sign of "protoindustry"[34]. Latent artisan skills related to self-sufficiency of farming families "would not have come to light, nor as forcefully as they did, if they had not been stimulated by the general upheaval caused by the Second World War [...], which witnessed town inhabitants evacuated to the countryside, the mechanical might of foreign armies transiting and fighting in the Region, the repatriation of soldiers who had seen the world, the pride of the womenfolk [...], the refusal by all to submit to isolation and subordination. The fifties [...] did the rest. When it became clear that young men and women rebuked the countryside and the rural homestead, where in any case [*ope legis*] the elderly remained as a rearguard, an exodus of men, of skills, of savings ensued: [...] this is where the proto (or as has been said the proteus) of Marche industry"[35] is to be found, supported by the dogged will of people accustomed to watching, imitating, learning, understanding, innovating, extending existing skills, assisted also by the adaptability of families to new types of work activities.

Marche hills. Photo by Gianni Berengo Gardin.

The individualism typical of the tenant farmer, "rooted, hard, tenacious, that in the farmer is stronger than in any other class, in which an extreme shyness and dark jealousy for the home emerge"[36] was thus combined with that of the artisan, giving rise to productive flows in which recent individualisation was and is expressed. In the central domestic continuity of the family, the urbanised tenant farmer operated and operates towards different choices borne from a threefold base: of origin, of the culture of work, of the determination to achieve productive differentiation in the common denominator.

Notes

1. *Inchiesta Agraria Jacini*, vol. XI, t. II, Rome, 1883, p. 756 of appendix to chapter XXI.

2. On the inquiry: A. Caracciolo, *L'inchiesta agraria Jacini*, Turin, 1958, and the introduction by G. Nenci to S. Jacini, *I risultati dell'Inchiesta Agraria*, Turin, 1976.

3. *Inchiesta,* cit., p. 757.

4. *Ibid.*, p. 758.

5. *Ibid.,* p. 757. The item "taxes" includes "meal", the "focatico" (hearth-tax), the duty on livestock, property wealth, etc. The food eaten with bread ("companatico") mentioned was almost always made up of dried fish (salted cod, stockfish) and beef, the former on religious vigils, the latter during festivities. On the diet and health conditions of tenant farmers, see S. Anselmi, "L'alimentazione dei contadini marchigiani negli atti della Inchiesta Jacini", in *Proposte e ricerche,* 11-12, 1983-1984, pp. 45-52, a twin file including other works on the history of food in the Marche Region; S. Anselmi, "Le condizioni fisiche dei contadini marchigiani nella Inchiesta Jacini", 1877-1884, in *Proposte e ricerche*, 16, 1986, pp. 68-68.

6. *Inchiesta,* cit., pp. 735, 736, 737.

7. *Ibid.,* pp. 734-764.

8. "The members depend for everything on the family head and owe their trust in everything [to the owner]" (Marone Pistagnesi), *ibid.*, p. 753; "submission and dependence in everything [of family members on the vergaro, as well as] obligation to every command of the owner" (Domenico Pieroni), *ibid.*, p. 751; "The [family] members must in all matters obey the head, who provides them with everything they need" (Giuseppe Navisse), *ibid.*, p. 749; "Total obedience and dependence [on the family head] and loyalty to the owner" (Elpidio Orso), *ibid.*, p. 748; "The family head presides over and directs all family affairs, on which all members depend" (Antonio Santandrea), *ibid.*, p. 744.

9. "Luxury, gambling, taverns and the smoking vice" (A. Santandrea), *ibid.*, p. 744; "The luxury of clothing, especially for women, gambling and drink for men" (Raffaele Marcantoni), *ibid.*, p. 746; "The luxury of clothing, especially in women, gambling [...]" (D. Pieroni), *ibid.*, p. 751; "The luxury of clothing" (Pasquale Tosoni), ibid., p. 755.

10. Tenant farmers generally speak of 10-12% "for usury" (*ibid.*, pp. 743, 745, 747, 751, 753); one claims that "you need a great amount of esteem to get it at 8%" (*ibid.*, p. 756); others state that the rates can cost up to 14-15% (*ibid.*, pp. 749, 754, 756), even reaching and exceeding 20%: "No less than 10%, but it's difficult to get it at that rate, it's often double that or even more" (*ibid.*, p. 762).

11. The tenant "class" "[...] needs to be protected against the theft [...] by day workers" (*ibid.*, p. 756); there is a need for "protection against country thieves" (*ibid.*, p. 758); "The [day labourers'] families always cause damage to the plants and the fruit" (*ibid.*, p. 754), "The most brazen [thieves] living in the tiny country houses are watched by the public authorities, but many are considered gentlemen and are even worse than the others" (*ibid.*, p. 763).

12. *Ibid.*, pp. 735, 736, 739, 751, 753, 755, 758.

13. *Ibid.*, p. 738.

14. The "casarolanti", on which much has been written, live by paying a "nolo" (rent) for a hovel normally located in a rural or suburban setting. They work on a day-by-day basis, according to what they find. Since they have lost their status and are no longer tenant farmers, other members of society – owners, artisans, bailiffs, public servants – wonder how they get by, implicitly accusing them of being thieves and moving them away from the city. Fearing theft, tenant farmers supervise them attentively, often forcing them to leave their lands, where even custom would otherwise authorise them to gather spontaneously-occurring edible herbs to allow them to "feed their families, and because this is what charity commands".

15. E. Sori, *L'emigrazione italiana dall'unità alla seconda guerra mondiale,* Bologna, 1979, p. 75, table 7.3.

16. E. Sori, "Crisi economica e crisi sociale: economia politica del crimine nella prima metà degli anni Ottanta", in S. Anselmi (edited by), *Nelle Marche centrali. Territorio, economia, società tra Medioevo e Novecento: l'area esinomisena*, 2 vols., Jesi, 1979, II, pp. 1.641-1.730. On the difficulties experienced by the Region during the transition to the new century, see also U. Tombesi, *Le condizioni economiche delle Marche,* Pesaro, 1905, and *La questione marchigiana*, Cagli, 1907.

17. Writes Giuseppe Tassinari in A. Serpieri, *Studi sui contratti a priori*, Bologna, 1920: "In the Marche Region, the movement [tenant protests] has spread just about everywhere, less in the province of Pesaro and Urbino, more in that of Macerata [in the Camerinese] and of Ascoli [near Fermo]" (p. 571), mainly due to the initiative of the Catholics, who demanded "greater retribution of the work coefficient, while respecting the rights of ownership", p. 272.

18. D. Spadorni, *Della mezzadria in relazione agli interessi dell'agricoltura*, Macerata, 1893, then republished with amendments and additions, under the title *Campagnie e Campagnoli nelle Marche*, Macerata, 1897.

19. L. Zoppi (edited by), *La Camera di Commercio, Industria e Agricoltura delle origini a oggi*, Ancona, 1962, pp. 209-210 and 214-215. The book illustrates a number of nominative cases.

20. On the farming contract and the family in the Marche Region (components, tasks, obligations, divisions, secessions, bookkeeping, etc.): "Capitolato colonico di mezzadria", in *Annunzi legali della Prefettura di Ancona*, n. 62, 1935, which enacts the arrangements laid down by the 1865 Civil Code, and those of the *Carta della Mezzadria*, 1933, with amendments; Inea, *Monografie di famiglie agricole*, XI: *Coloni e mezzadri della Marca di Ancona*, Rome, 1935; B. Cioffi, *Il volto agricolo delle Marche*, Bologna, 1953; A. Ciani, *Usi e consuetudini in materia agricola vigenti nella Provincia di Ancona*, Ancona, 1956. As regards the tenant farming family in the Marche, see the stud-

ies by C. Verducci, "La famiglia colonica marchigiana nel XIX secolo: note sul Fermano", in *Proposte e ricerche*, 2, 1978, pp. 101-123; A. Mosca Proietti e E. Insabato, "La famiglia mezzadrile marchigiana fra tradizione e mutamento", in *Proposte e ricerche*, 5, 1980, pp. 5-22; A. Cortese, "Modifica della famiglia marchigiana attraverso i censimenti", 1861-1981, in *Proposte e ricerche*, 16, 1986, pp. 82-90; C. Verducci, "Strategie e dinamiche familiari", in S. Anselmi, *Marche*, *Storia d'Italia, Le Regioni*, Einaudi, Turin, 1987, pp. 453-473.

21. B. Ciaffi, *op. cit.*, p. 181.

22. Inea, *op. cit.*, p. 39, refers back (with slight adjustments) on p. 17 to "use coefficients by gender and age" by Arrigo Serpieri, republished in *Istituzioni di economia agraria*, Bologna, 19563, and adopted by B. Ciaffi, *op. cit.*, p. 180.

23. B. Ciaffi, *op. cit.*, pp. 767-771.

24. B. Ciaffi, *op. cit.*, p. 185. See also for the long period A. Palombarini, "Gli abiti dei contadini e dei popolari nell'Ottocento marchigiano", in S. Anselmi (edited by), *Contadini marchigiani del primo Ottocento. Una inchiesta del Regno Italico*, Senigallia, 1995, pp. 57-96, which states that luxury is an exception (p. 61), that jewellery was sometimes made with fake coral (p. 63), that the dispute over luxury was due to the fear of changes to customs (early 20th century), because "the assets of peasants are limited to the days of solemn festivities", p. 61.

25. G. Porisini, "Produttività e agricoltura: i rendimenti del frumento in Italia dal 1815 al 1922", in *Archivio economico dell'unificazione italiana*, series II, vol. XVII, Turin, 1971, p. 251 in appendix.

26. S. Anselmi (edited by), *Inserimenti rurali, case coloniche, economia del podere nella storia dell'agricoltura marchigiana*, Jesi, 1986. See also A. Mori, *La casa rurale nelle Marche settentrionali*, Florence, 1946; L. Brigidi, *La casa rurale nelle Marche centrali*, Florence, 1953; Inea, *op. cit.*; B. Ciaffi, *op. cit.*, pp. 111-150; M. Dean, "Problemi di cartografazione della tipologia abitativa rurale nelle Marche", 1930-1950, in *Proposte e ricerche*, 16, 1986, pp. 91-103.

27. B. Ciaffi, *op. cit.*, p. 185.

28. S. Anselmi, "Padroni e contadini", cit. : "The boom in savings was connected to the price of wheat which, after the peak in 1926 (200 lire per hundredweight) and the slump of 1927-1934 (from 140 to 86 lire), started rising again as of 1935 through to 1947, following a trend that, though largely distorted by inflation, brought prices up from 105 to 13,850 lire, with an almost constant growth on relatively stable values between 1935 and 1942. The price of wheat drove up that of other cereals, as well as of wine, oil, eggs, potatoes, livestock" (pp. 289-290).

29. *Ibid.*: "Once again, the war marked the triumph of women and the elderly, performing to the depths of the countryside its terrible accelerating function, then centralised by the return of the men, who were far different from those who had departed a few years earlier" (p. 290).

30. W. Kula-J. Kochanowicz, "Contadini", in *Enciclopedia Einaudi*, Turin, III, 1978, pp. 901-934., p. 928.

31. Inea, *op. cit.*, p. 39.

32. See entire passage: *ibid.*, p. 41.

33. For a thorough reading of the domestic life of a farming family in Osimano, see pp. 39-62 of the Inea monography, cit., in particular p. 40 for the quote.

34. See for the emphasis on the European phenomenon n. 59, 1985 of *Quaderni storici*, edited by Carlo Pani, pp. 341-528, and n. 23,1989 of *Proposte e ricerche*, edited by Marco Moroni, for the Marche Region, pp. 59-186, and for the free use in Italy of theories developed elsewhere, the observations made by S. Anselmi on pp. 180-186.

35. S. Anselmi, "Il dibattito sulla protoindustria", in *Proposte e ricerche*, 23, 1989, p. 185.

36. B. Ciaffi, *op. cit.*, p. 69.

The landscapes of the Marche

Armando Ginesi

What is the environment, in particular the natural one, i.e. the landscape? Does it have a relationship with man's ways of being and behaving? Can art be influenced by it? If so, what is the connection, in the reality of the Marche Region, that binds the environment, *ethos* and visual art?

The term environment has a number of meanings. The linguist Giacomo Devoto gives us a few: "The surrounding space considered with all or most of its characteristics"; "The combination of physical, chemical and biological conditions that allow and favour the life of living beings"; in the figurative sense: "Complex of social, cultural and moral conditions in which a person is found, formed, defined".

Given that the noun "environment" ("ambiente" in Italian) derives from the present participle of the Latin *ambīre* – meaning to "go around" – its most appropriate sense ends up being that of "everything that surrounds".

In the specifically artistic context, the idea of environment was defined by Leon Battista Alberti, in *De re aedificatoria* in 1452, as "place that surrounds buildings". The great Humanist referred to architectural art, inside a culture of urban planning that was being reborn after centuries of oblivion and that in him found one of its first and most authoritative exponents. Immediately thereafter, identified also with the meaning of the Latin term Region, the concept of environment was broadened to denote not only the place that surrounded buildings, but also, and above all, that space surrounding individuals and their way of being. Environment was beginning to be understood as something that surrounds the person, like nature, the landscape, the city, the social context.

When placed in relation to the artist and the artist's activities, the environment becomes a complex of nature, society and culture that embraces and expresses. In an era not too distant from our own, this idea was put forward by Hippolyte Taine in *La Philosophie de l'Art* (1865), in which he formulated the theory of the *milieu* to define the nature of a work of art. He pointed out how the artist does not live in isolation, but is in fact an integral part of a system characterised by a given level of civilisation, of specific behaviour patterns, by a certain type of ideality that acts as a determining cause of art as it is produced in a precise period of history, in a particular territorial environment, inside a specific socio-cultural group. By these theories, Taine virtually conducted the first sociological studies of the artistic phenomenon.

There can be no doubt that it is the environment – construed in its totality, therefore as "everything that surrounds" – that intimately bonds with the personality of the artists with whom it establishes a relationship of mutual influence by which sensations, stimuli, emotions and ideas that by osmosis transit in one direction or another. Nor can there be any doubt that, of the environment in general, the feature of greatest interest to the artist was (and still is) that of nature, i.e. the landscape, the vision of the original *habitat*; in a certain way the primary *òikos* where the existential circumstance of man was born and developed.

The artist of any era (from the remote times of the Palaeolithic, an era in which the first forms of "artistic" representation made their appearance, to contemporary times) has always deeply perceived the natural environment, of which they recognise themselves as an integral part, the carriers of "stigmata", the inspirational element, direct or indirect, of their creativity. Even when it did not constitute a specific theme of current interest.

With special reference to the visual arts, it should be noted that painting, being the most suited to visualising relations with the natural environment, has always

Mechanisation of Marchegian industry.

construed the landscape as a value of enormous importance by which the *Weltanschauung* and *Kunstwollen* of different civilisations and eras have been depicted. This can be clearly observed in the study of art history of the West as much as of the East.
While remaining in our western context, it might be recalled that, throughout the Middle Ages until the proto-Humanism of Giotto, the landscape has been considered the symbolic vision of nature, while during the Renaissance it was seen from a perspective and realistic manner.
Some artists depicted it with idealism (Giorgione, Titian, the Carracci, Domenichino, Poussin), while others opened up to fantastic solutions (Altdorfer, Grünewald, Salvator Rosa). Yet others preferred a more faithful approach by adhering to the naturalistic model (Constable, Corot, Courbet, the Barbizon school), interpreting it through stimulus deriving from fleeting impressions (the Impressionists), or representing it under the influence of theories on colour and optic perception formulated by Chevreul, Helmholtz, Henry, Rood (the neo-impressionists Seurat and Signac).
Paul Cézanne conceived the landscape as a rational grid structure onto which multiple sensations could be positioned in order, while even the historic avant-garde of the first quarter of the 20th century, by their disturbingly innovative poetics, presupposed its presence in all expressive hypotheses. The same aniconic art (improperly defined as "abstract"), though not treating it as a theme to be visualised, almost always considered it a value to be reduced to its most essential components (Piet Mondrian by his series of increasingly bare trees, from *Red tree* of 1909-10 to *Gray tree* of 1911 and *Apple blossom* of 1912) or assumed it as a dimension of the conscience, excited by the existentialist philosophy (Fautrier, Tàpies, Burri, Millares, Saura, Guevara, with informal matter art of the fifties and early sixties).
Without a shadow of a doubt, the natural environment and its constituent parts (tree, mountain, sea, river, panorama and so on), beyond their visibly objectivised presence, have always contributed to the determination of the temperament of artists, to the formation of their ethos, and thus to the very genesis of their poetics.

The Marche: a "distillate of Italy"

Almost at the centre of the Italian peninsula lies the Marche, a Region of hardly 10,000 square kilometres and less than one and a half million inhabitants. The plural form of its name in Italian tells us that it is an articulate, multifaceted territory, though unique in many respects; its parts always melt together in a dimension of synthesis that represents the peculiarity of a land that Guido Piovene defined in his *Viaggio in Italia*, edited between 1953 and 1956, as the "most classical in Italy".
This journalist and writer from the Veneto composed many precise pages on the landscapes of the Region and the characteristics of its inhabitants, showing an ability to capture its essence perfectly.
"The Marche are a plural", writes Piovene, "the north has a tinge of Romagna; the Tuscan and Umbrian influence is apparent along the Apennine range; the province of Ascoli Piceno is the antechamber to the Abruzzo and the Sabina. Ancona, maritime city, is a part for itself [...]. But even though they reflect them, the Marche do not really resemble either Tuscany, or Romagna, and not even

Osvaldo Licini, *Paesaggio verde cupo* (1926). Turin, private collection.

the Abruzzo nor Umbria". As regards this typicality due to the synthesis that the Region represents, the author refers to the hillside, stating that in the Marche it "is almost a large and natural Italian garden. It is neither the Tuscan hillside, nor the Umbrian, nor the Veneto. It is gentle, serene, pathetic, lucid, without sharpness [...]. The hills are rounded, with long, slow, grassy slopes, sprinkled here and there with large solitary trees [...]. It is the prototype of the idyllic, pastoral landscape". He then continues: "If I were to establish which is the most typical of the Italian landscape, I would have to say the Marche [...]. Italy is overall a kind of prism in which all the landscapes of the earth are reflected, present in moderate proportions, one blending into the other. With its landscapes, Italy is a distillate of the world, and the Marche of Italy. Here we have the most complete example of that average, gentle landscape without softness, balanced, moderate, almost as if man himself had drawn it".

This insistence on the value of the synthesis highlights the classical nature of a territory that, with its medium nature, forges, as will be seen later, the balanced temperament of its inhabitants, leading them towards a reflective approach which, in turn, plays a decisive role in defining the distinctive features of artistic expression. "In the Marche," continues Piovene, "you won't find the primitive or the extremely modern. Nothing exaggerated. It is a filtered, civilised land, indeed the most classical of our lands".

But is the Marche landscape actually how Piovene sees and depicts it? Michele Dean, in *Storia d'Italia: Le Regioni dall'Unità ad oggi, Le Marche,* published by Einaudi in 1987, gives us a scientific description of it: "The configuration of environmental framework of the Marche, as prospected in the intensely elaborated landscapes of the Region, owes its origins, structural characteristics and therefore its way of unfolding, to the tectonics and orogenesis of the Apennines, while successive modifications derive from the modelling action of exogenous agents [...] in particular [...] the constructive – and destructive – action of men, who have always acted on the territory, organising variously-sized portions of it, in the forms corresponding to the different stages of their history".

Besides the technical explanation – which underscores the human impact on the landscape, indicating how the character of its inhabitants is not unalike the balanced traits bestowed by nature on the Marche – historic testimonies tend to corroborate the writings of Piovene and other contemporary writers and journalists. Take the Medieval poet Cecco d'Ascoli, who spoke of the Region as the "bel Paese de li dolci colli" , or the chronicles of ambassadors and travellers of the 16th, 17th and 18th centuries who constantly describe the "beauty of the fields", the "tidy fields" and so on.

Nando Cecini, in the *Storia* by Einaudi quoted above, penned an essay entitled "*Le Marche: una metafora per i viaggiatori*" in which he reports that the territory, in the eyes of the foreign traveller, is "an ideal grid in reality-fantasy relations" as well as "a literary metaphor. A Region that is well suited to this balanced operation between landscape and civilisation, between nature and anthropisation. A miniature ecosystem: from the ruggedness of the Apennines to the open fields, the gorges furrowed by torrents to the harmonious stretches of the waterfront. Besides the landscape, the presence of a secular civilisation, gathered in the historic microcosm of abundant urban centres".

What Cecini is describing are the deductions of the Marche environment made by foreign travellers in historic times. Almost all spoke of hills and brief fertile plains covered with trees and vineyards: these include Montaigne (*Voyage en*

Three trees by Piet Mondrian.
Top to bottom: *Red tree* (1909-10),
Gray tree (1911), *Apple blossom* (c. 1912).
The Hague, Gemeentemuseum.

Italie), Dionisio Atanagi, the Bolognese Dominican Leandro Alberti who, in *Descrittione di tutta Italia*, published in Bologna in 1550, described "pleasant and delightful fields", hills, plains and gardens that evoke a return to Eden, the earthly paradise lost.

Leonardo da Vinci himself, visiting Pesaro and Urbino in 1502, exalts the beauty of the Region, which he defines a "proportionata armonia". And despite his reputation as a disillusioned, morose writer, Maximilian Misson, in the *Nouveau voyage d'Italie avec un Mémoir contenant des avis utiles à ceux qui voudront faire le mesme voyage*, edited in La Haye in 1691, cannot help but use a similar descriptive tone.

Numerous other testimonies can be cited, including those by Charles de Brosses, Denis Ivanovic Fonvizin, Stendhal, Johann G. Seume, Wolfgang Goethe (who defined the Marche "a wonderful piece of all creation"), Madame de Staël, Hans Christian Andersen, André Gide.

These glowing descriptions of the Marche landscape have thus continued through the centuries – seen as something serene, flourishing, beautiful, measured and harmonious. This would indicate that, at least as of the Middle Ages, the balanced characteristic of this land has remained structurally unaltered, to the point that in modern times, as we have seen, Piovene can safely claim that no exaggeration exists in the Region, in perfect agreement therefore with another author (and authority on landscapes) A. Sestini, who in *Paesaggio* (Milan 1963) states that, "the most common aspect of the mountain landscape in the Marche is the broad grassy tops, almost flattened and rounded, or the impressive domes that overhang steep flanks".

When quoting Piovene, we mentioned that he noted the "average, gentle landscape without softness, balanced, moderate", almost as if drawn by a human hand. A reference to the relationship of nature and man as if to a *tòpos* in which influences are mutually exerted.

The concept is taken up by a researcher of the Marche environment: Gianluigi Mazzufferi, who in the chapter "L'ambiente naturale" in the volume *Nelle Marche Centrali* (edited by Sergio Anselmi and published by the Cassa di Risparmio di Jesi in 1979) writes: "However massive, destructive and predatory, the impact of man on the environment over almost twenty centuries still accounts for a primordial agricultural structure and a plant heritage that cannot be damaged beyond certain limits". Limits that are not exceeded because, as Mazzufferi admits, the inhabitants have "respected traditional canons of behaviour".

Aurea mediocritas?

There are thus close interrelations between environment and behaviour; the latter is largely built up under the influence of the former, while the former gradually takes shape under the impact of the latter. For example, the conduct of mountain communities differs to that of valley and seaside communities, while the environmental differences produce cultural diversity. At the same time, each culture produces its own way of interpreting and inhabiting the native environment in which it continues to evolve, developing typical models of intervention to transform the environment according to its own ideas.

As regards the behaviour of the people of the Marche (i.e. the complex of thoughts and consequent actions that go under the name of ethos), the harmony with balance, the sobriety but also the solidity of the surrounding landscape seems clear. As early as the late 18th century, Amedeo Crivellucci, in *Una comune delle Marche nel 1798 e 99 e il brigante Sciabolone* (in the edition consulted, published in Pisa in 1893) wrote that the people of the Marche were not "accustomed to worrying about anything but the rain and the fine weather", bent over the ground that "they water with their own sweat [...] fatten of themselves [...] remove [...] knead [...] embrace [...] love [...] like a mother, like a bride and for whom they live and die of hunger and poverty".

In his statistics reports compiled in the years 1864 and 1881 on the conditions in the various provinces of the Marche, the prefect G. Scelzi described the people of the countryside "active and sober" several times. Patrizia Sabbatucci Severini, in the chapter "L'aurea mediocritas: Le Marche attraverso le statistiche, le inchieste e il dibattito politico-economico" in the *Storia* published by Einaudi and mentioned earlier, refers back to the report on industries in the province of Ancona (published in 1861 at the request of the organising committee of the first national exhibition), in which the method of tenant farming – typical of the agricultural economic model of the Marche – is described as one of the causes for the delay in the process of modification of entrenched methods which nonetheless "make the farmer fond of the earth and doubles his labour". This provides us with the explanation of why the Marche peasant feels such a bond with the earth, which he loves and to which he totally devotes the meaning of his existence, condensed into and symbolised by work.

Sabbatucci Severini also quotes the inquiry conducted by Ghino Valenti, the main editor of the texts in the Marche report in the said *Inchiesta Jacini*, stating that he "summarises the state of agriculture and Marche farmers in the motto *aurea mediocritas*, where the latter term refers to the intermediate degree of intensity of crops, and the former to the "relatively happy [condition] of the great majority of farmers".

Furthermore, Sabbatucci Severini writes that "tenant farming and artisan crafts have an analogous effect on the economy and the public spirit, forging a sense of community and specific, peculiar moral characteristics: individualism, clever industriousness, hardness of sentiment (shown to family members through the capacity to work), mediocrity of aspirations (which brings people of every standing and class together), egalitarianism".

Tenant farming and artisans produced a culture and specific modes of life: an attachment to the earth and materials, the assets constituting the reason for survival; interest in work, because the more the work, the greater the well-being; the duty and pleasure of doing, and above all of doing it well; gratification in seeing the results of labour and commitment (farming products, crafts), from which the quality of life derives.

Besides generating a strong ethical sense for work (and especially work well done), all of the above helped develop a culture of invention manifested in the on-going aesthetic refinement in crafts and the decoration of the farming cart (the *biroccio*, from the Latin "*bi-rota*", "with two wheels").

By combining his practical and utilitarian sense with the pleasure derived from making surrounding objects and spaces more attractive, the peasant worker then the artisan progressively developed a highly refined sensibility for handcraft design, which itself was one of the main reasons behind quality production (or at least an interest in quality) that is a hallmark of the industrial enterprises of the Marche Region.

Another manifestation of the innate sense of beauty typical of rural peoples is the remarkable attention paid to crops. According to written accounts, the fields have had their current-day appearance ever since ancient times: carefully tended, almost combed. By the sharing of property and rotation of crops, the arrangement of trees, vines and sowed land that form perfect geometric shapes and colour combinations of green, copper, brown and yellow are redolent with poetic connotation.

The ethical process of *doing-doing well-invention-aesthetic sensitivity* gave rise to a sense of dignity of the family, seen as a strong aggregate of individuality, a way of being together while being each member different; a model of synthesis on which the very identity of the Marche was moulded: a homogenous, unified plurality. As much is pointed out by Sergio Anselmi, refined researcher of the economic history of the Region and editor of the much-quoted *Storia d'Italia* by Einaudi. In his preface to this volume, after recalling that, following national unification, the Marche saw its six legations of Urbino, Ancona, Macerata, Camerino, Fermo and Ascoli Piceno reduced to four provinces (Ancona, Pesaro, Macerata, Ascoli Piceno), Anselmi writes that they constitute "a small rectangle of a peninsula, quite taciturn, full of farmers, the clergy, land owners, traders and transporters of wheat and corn, thick with hills and tiny communes, well cultivated and a little coarse". But above all he states that "the birth of the Region as an organisational body in the late 1960s ratifies the Marche, which remains a plural in name while tending towards homogeneity".

Anselmi also rebuts the historic label of *mediocritas* obstinately applied to the Region, as recalled by Sabbatucci Severini, "formulated by the surveyors of the *Inchiesta Jacini*, as well as by a number of researchers, politicians and critics of the Marche between the late 19th century and the Second World War". We concur with Anselmi that the use of this expression is erroneous, preferring to substitute it with 'medium', in the Aristotelian sense of the happy medium between extremities, and consequently of the synthesis that Guido Piovene defined and often recalled as environmental and landscape connotation perfectly reflected in the nature of the Marche people (their "marchigianità").

The judgement of mediocrity – ultimately negative because reductive – probably derives from the fact that, as our journalist and writer from Vicenza with his usual insight points out, the Marche is not a popular Region. He wrote that "the popularity of the Marche is perhaps undermined by the absence of the extravagant, the surprising, the exciting things that attract the attention of those who seek the extraordinary".

Exactly: the Marche landscape lacks excesses and superlatives as much as its inhabitants, whose behaviour (even in politics) reflects the paradigm of plurality that becomes homogenous through synthesis. This is noted by Piovene, referring to situations personally experienced and recorded in the late 1950s: "In politics, the Marche tend towards red in the province of Pesaro, where the thrust of Emilia and Romagna runs out; divided in the province of Ancona, where the Christian Democrats however already prevail; while the provinces of Ascoli Piceno and Macerata are Christian-democrats, or rather governmental".

At this stage we have a plurality that gives rise to a conclusion that is oriented, also ideologically with practical consequences, towards synthesis and the value (for Aristotle a moral virtue) of the medium: "Whatever the political opinion, it is never poisoned by extremism of character. The Marche does not draw attention for its wealth, nor for its poverty. There are no lands that require substantial reclaiming; they do not host enormous masses of unemployed workers. The

Female workforce in traditional industry.

tendency is actually to go about their own business [...]. Dedicated to farming, this Region is prevalently a vast republic of country folk, tenant farmers and small land owners, looking after practical problems, jealous of their privacy".
The world described by Piovene, as earlier stated, is that of the Marche during the fifties. Today the productive reality has changed: for example, the abolition of tenant farming in 1964 heightened the role of small business, which was transformed in many cases into small and medium-sized industry. However the basic culture, the stability, balance and moderation, the loved tenet of "biting off no more than you can chew", have all remained intact in the thinking and being of the people.
It is "difficult to find elsewhere such an exact match between mind and landscape", writes Piovene, thus strengthening the idea of a close relationship between environment and *ethos*. He then adds that the character of the typical Marche inhabitant "is perhaps without clear qualities, but is precisely by this strongly defined: an average Italian character" (the return of the sense of the medium and multiplicity condensed into the unitary) "that strikes a balance between discordant experiences through compromises, the diplomacy of common sense. The Marche is a crossroads between South and North, where the characteristics of the Italian come together and are attenuated. A people like this, of composite nature, are perhaps the most simple and homogenous in Italy".
Returning to what has already been mentioned about the people's commitment to work, their interest in "doing things properly", their sobriety, we quote another passage from Piovene in which he asserts that the Marche inhabitant is "a hard worker, without excess, because he prefers a modest life; a subtle, ironic intelligence, that convinces him to be happy with little[.] The average family conducts a sober, composed, uneventful, slightly staid life".
The writer has seen correctly: the downside of this type of self-sufficiency is the risk of stagnation, a tendency to maintain the status quo ante. However the positive aspect of a substantially autarchic approach, still strongly influenced by a distant economic model that prioritised self-sufficiency, no waste and recycling, has been the constant stimulus to inventiveness, creativity, intelligence in general, not to mention a refined sense of irony perceived by Piovene and thanks to which negative experiences are tempered and expelled from the mind.

Marche art: solitude without isolation

When one lives sufficiently satisfied inside a personal vital space, with an ethic (almost Calvinist in its radical nature) of doing and doing properly, a tendency inevitably develops towards aesthetic activities, to beautify one's *habitat* and the objects that surround it, starting with the work tools through to the home.
The process, which originated in prehistoric times (in particular during the Neolithic era anticipated by the Mesolithic), when the groups gave up their nomadic lifestyle after discovering the rural economy of harvesting (instead of hunting and gathering), gave rise to the first, rudimentary types of design in the crafting of useful yet attractive objects, of decorative painting and sculpting.
Almost all farming societies – as in the case of the Marche society until a few decades ago, though retaining the cultural foundations, passed down initially to the crafts trade, then to small and medium-sized industry – have undergone the above process, on the basis of the original Neolithic paradigm, explaining why aesthetic sensitivity is so widespread throughout the Marche Region.

Two paintings by Osvaldo Licini.
Above: *Amalassunta su fondo rosa antico*, 1950.
Below: *Amalassunta gialla*, 1950.
Ascoli Piceno, private collection.

Previous pages:
Marchegian landscape
by Gianni Berengo Gardin.

The reasons for the close relationship between landscape and the people, between nature and *ethos* also help to clarify the bond between creativity and another two factors, given that art is so deeply related to both the environment and the complex of thoughts and actions that inhabit the context and produce art. In short, what Taine had claimed by his above-mentioned "theory of the *milieu*". His thoughts can also be integrated with the ideas of other successive researchers, such as Alois Riegl and Wilhelm Worringer. Riegl maintained that the symbologies implicit in ornamental morphology always reflect the intuition of the time and space of a specific ethnic group and dependent on the type of experiences had. Worringer, in 1906, however theorised on representative shapes construed as the manifestation of the consciousness of a reality that constitutes the model. He then made examples to show how the Mediterranean peoples produced classical art (thus bound to natural values) to imitate a nature seen positively, while Nordic peoples produced abstract art to reflect the tensions and conflictuality between man and the environment, perceived as indefinite, hostile and treacherous.

All contentions are worthy of interest, provided that they are not taken in absolute terms or considered exhaustive, lest we run the risk indicated by Giulio Carlo Argan: the "fixation of national or ethnic formal constants and the subsequent hardening of the historic process". Indeed "if closer and more frequent relations between phenomena of the same cultural area cannot be denied, this is not a matter of invariable constants but of historic developments".

From these general principles, let us now move on to the reality, by reading another extract from Piovene on the relationship between the Marche and the writings of Leopardi: "Between Leopardi's verses and passages, one can glimpse against the light not only the landscape of the Marche, but also the character of the men. In him resides, sublime, a supreme prudence of intellect, a refusal of ideas and events that are called progressive, a conviction that the nature of man is unchanging, a mistrust in the metropolitan, foreign and in general other fashions, typical of classical provinces and noble blood".

To avoid misunderstanding, it should be clarified that we do not believe in the existence of a specifically "Marche style" in art. The issue was long debated in 1974 during an important exhibition, supervised by Carlo Antognini and organised by the Amministrazione provinciale di Ancona in the palazzo Pianetti of Jesi, entitled "*Marche Arte 74*", by which a critical reflection on the contribution made by 20th century artists hailing from the Marche Region was initiated. Antognini, an academic particularly attentive to contemporary Marche poetry and visual art, had maintained that the daily activities with the sea and the fields had made the Marche people "taciturn, secluded, shy towards easy openings, yet more inclined towards melancholy than sadness, wondering than anguish", specifying that his place of wondering was becoming "the memory which is asked questions and gives answers, everything veiled by a nostalgic mood that filters and mediates, expanding it to include a vast sentiment of melancholy". He had also written that the Marche artist "harbours, side by side and without contradiction, the impetus of the great and the judgement of the limit, the aspiration towards classicism and the need for abstraction, the venting of the senses and the *ictus* of intelligence; above all, a profound sense of measure, combined with a natural scepticism, which sharpens the assessment of what is essential and the fleeting nature of the moment or the pragmatic formulas, making him concretely an avant-garde, and never an experimental, artist".

Antognini had then identified a "Marche artistic line" as a consequence of shared ethnic and behavioural characteristics of artists: "It seems to us that the most acute souls and minds, that is those artists who most and best have attempted to artistically depict these basic qualities and this intellectual and metaphysical tension [...] all together indicate, precisely, what might be defined, without excessive presumption, an artistic line of the Marche; determined by the sum of values and the psychological conditions explained so far, but also by a combination of common reflections and replies, though in different creative openings, to fundamental existential questions that today's artist asks more insistently than those of the past".
Enrico Crispolti opened a debate on the question when he contested the thesis of a "Marche artistic line", i.e. the existence of a style that was characterised to the extent of being distinctive from the expressive trends witnessed nationally. We also disagree with the identification of a specific "Marche line", if it implies a system of linguistic and formal modules culminating in a clear expressive direction as described by the formalists, Riegl and Worringer. We therefore believe that Crispolti is right; though so is Antognini, because he had no intention of claiming the existence amidst Marche artists of a common grammatical, syntactic and lexical system, but rather a single matrix of culture and *ethos* – in short, a common "temperature".
In conclusion, while acknowledging in all Marche artists a cultural and temperamental uniqueness, the typical plurality of the Marche remains clear and intact, a kind of prism in which all the landscapes and moods of Italy are reflected – after having been composed and unified in a synthesis of ethical values based on common sense, humility, industriousness, moderation – and reproduced in art, then expanding in all directions thanks precisely to the multiplicity of expressive and linguistic temperaments.
The Marche thus represents a screen on which all lines of research conducted nationally and internationally can be projected, demonstrating that the Region – despite its median position making it an element of friction (thus rather compressed) between the developed North and the repressed South, the peasant origin of its culture, and the shyness of its inhabitants – is not culturally nor artistically isolated. Indeed its artists, from the most celebrated to the more modest, have made their presence felt. They may be solitary, but certainly not isolated.
The average inhabitant of the Marche has not inherited the negative aspects typical of the peasant culture, such as chronic diffidence, backwardness, isolation. In fact (and what underscores his peculiarity) he is open to the world beyond the boundaries, because he is curious, prepared to consider – attentive yet cautiously, though not mistrustful or hostile – other cultural models. His antennas are outspread to pick up whatever is new and unusual.
His distinctive nature surfaces when these values, foreign to his own direct life and cultural experience, arise; he studies and analyses them, slowly, meticulously, thoroughly – not a cursory glance, but a painstaking survey. Like the peasant when he goes to market, before buying the object he turns it over continuously in his hands, feeling and shaking it, as if to become acquainted with it by his senses. Only at the end of a long and thorough sensorial examination does he decide whether to buy or look elsewhere.
With the very same circumspection that a peasant inspects and assesses an object for the home or work, the artist and the poet of the Marche confront the prob-

lem of the relationship between self and the outside world. They are informed on the artistic issues of their time, follow the debates, experience the contradictions then transfer the results to their own workshops where, after having donned the artisan's apron, turn the data and information collected over and over again, shaking, feeling then finally processing them into original language forms.

The Marche Region bristles with this type of workshop, in which shy and retiring artists work tenaciously and with great dedication, as already mentioned solitary but not isolated. It is also for this reason that, despite its relatively small territory and population, the Marche has always made a leading contribution to the development of Italian, European and international art. The phenomenon has been constantly repeated throughout no less than the last six centuries, with periods of great intensity, such as has been witnessed this century.

The trend began right back in the 14th century (when the people from the Marche were called, in particular in Tuscany, the "Marchiani"), especially in the areas contiguous to Umbria where local quality artists were active, dispersed with those that contractors summoned from the near north, especially from Rimini and surrounding areas. It continued through the 15th century, with indigenous artists flanked by renowned foreign artisans invited by religious brotherhoods, noble families, diocesan curias, monastic orders and convents, rich merchants who identified in their expressive modules the features corresponding to their own nature and culture.

But it was the 16th century that witnessed the flourishing of authentic champions of the international art world, to be joined by others from abroad, drawn either by contractors interested in cultural values of whatever origin, provided they shared an appreciation for the moderation, equilibrium and simplicity of Marche civilisation, or by the fascination for the quietly beautiful natural environment complemented by an enviable quality of life.

By taking the example of the 16th century, it is possible to make a paradigmatic definition of the three ways in which the Marche played a vital role in the evolution of art. This took place (and still does) by means of three types of artist: the artists who were born and remained; expatriate artists; and the artists who, though not being native residents, came by choice or necessity to the Region, discovering there an ideal cultural and existential tòpos.

The first group was represented by a number of important names, including Timoteo Viti, Girolamo Genga, Raffaellino Del Colle and various anonymous artists; the second, by Raffaello Sanzio and Donato Bramante, to mention only the greatest; the third included an impressive series of celebrities, from Perugino to the Vivarini, Giovanni Bellini to Carpaccio, Titian, Savoldo and Sebastiano Del Piombo through to Tintoretto and Andrea del Sarto. But probably the most emblematic figure of this artist type, attracted by the Marche, was Lorenzo Lotto.

The three categories of artists bound to the culture of the Marche Region were they themselves subtly yet intensely interconnected by a tie that has defied the centuries. The reasons for this phenomenon are plentiful, but the most significant is certainly the common denominator of "memory", the founding value of Marche culture, and in particular of its artistic and visual culture.

The painter and the sculptor of the Marche Region (as much as the writer and the poet) are like Ulysses: they build their life projects and poetics on the mnemonic dimension, used as an instrument of transfer "into the heart of ancient events", as noted by Luigi Martellini in his essay *La poesia delle Marche*.

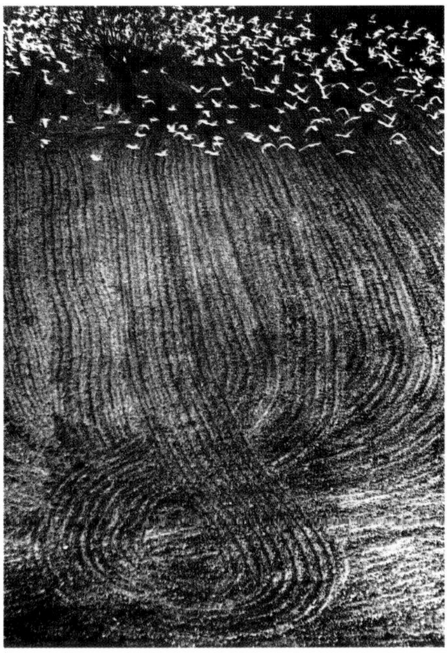

Mario Giacomelli,
Sistemi del paesaggio (1980-94).

This attitude is even further accentuated in the expatriate artist, whose memory is also used to preserve an inner image of the native land, that like for Ulysses acts as a stimulus, in the production of nostalgia.
As regards the group of non-indigenous artists who remained fascinated by the beauty of the Region, it is worth recalling the definition that Piovene gave to Italy – a prism in which the landscapes of the world are harmonised and reflected – and that of the Marche, a prism of Italy. This provides a certain explanation of why every artist from outside, whether Italian or foreign, ends up finding in the prism of the prism one or more features analogous to his or her own identity, able to 'couple' the artist to the local *ethos*, and consequently to introduce him or her into that balance achieved between landscape and sentiment that is the hallmark of the classicism of the Marche geography as much as of its soul.
The artistic-visual poetics of the Marche – native and adoptive – people are built on the values of memory, not to mention the sense of balance and serenity that the landscape imbues. Even the classicism of Raffaello and Bramante was certainly derived from this sub-layer of natural classicism and the nostalgic memory of it: do not the rounded domes of Bramante's architecture recall the description quoted from Sestini, who claimed that the most common feature of the mountain landscape of the Marche was that of broad grassy tops, almost flattened and rounded, or the impressive domes...?
Luigi Martellini recognised earlier this peculiar trait in the poet of the Marche (the clearest, most illustrious example of the 19th was certainly Giacomo Leopardi); referring back to the platonic concept of memory as an instrument most suited to transferring man into the heart of ancient events, he specifies that "as regards the Marche poet, it is important to identify the concept of Plato recalled above with the more complex one of conservation, which is nothing but the materialisation of the intellect and oneirism which, through the filter of a kind of poetic *furore*, is transformed for certain people into notion of invention".
The inclination towards reminiscing is encouraged by the harmonious and classical nature of the landscape, so widely discussed earlier, which has always culminated in the formation of a stable, solid temperament. The lack of the exceptional, the hyperbolic and the surprising has forged a meditative, introspective character; the external world is introduced into the ego by the mechanism of nostalgia and the return to the past, towards the roots of one's own history, individual and collective, ontogenetic and phylogenetic.
At this stage it is necessary to recall Antognini once again, as he emphasised the inclination of the Marche people towards melancholy and interrogation, identifying in memory the place allocated to questioning, with nostalgia that filters and mediates, by melancholy, both questions and answers.
All Marche artists – even those taking the path of design – produce poetics charged with mnemonic significance. There have been both past and contemporary authors who have travelled the way of a design culture, drawn by the rational and the idea of calculation (sublime examples of mathematical and proportional painting were left in the Marche during the 15th century by Piero della Francesca), though only after having come to terms with the moods of reminiscence and nostalgia, used to warm the structure of rationality. Argan would have said that they translated the rational into ideation – which we might add is still a form of imagination.
A remarkable example for all is Osvaldo Licini, who demonstrated to the contemporary world that poetry is possible through geometry, by recovering the

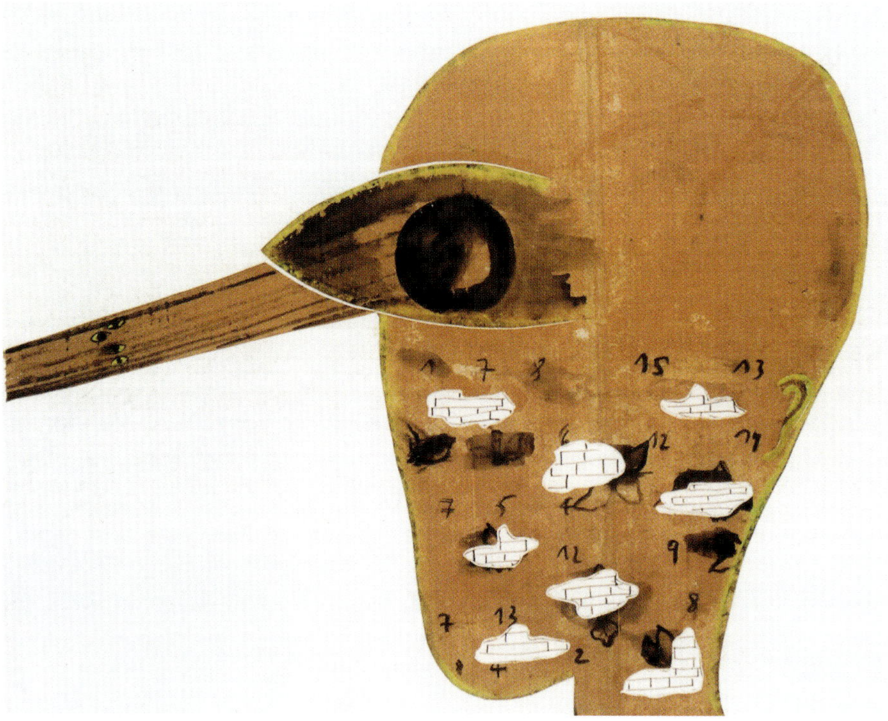

Two paintings by Enzo Cucchi.
Above: *Memoria*, 1996.
Below: *Luce tra cielo e terra*, 1996.

thought of Pythagoras, who had created a poem, in fact a mysticism, of the number, a religion that exalts, able to enlighten the soul of measurement and calculus.
An ethos of doing and doing well; inventive intelligence; sobriety, moderation and balance; solitude but not isolation; memory, introspection and nostalgia are the most significant traits of the inhabitant and the artist of the Marche, immersed in that culture of the Leopardi "hedge", the memory that did not however prevent the great poet of Recanati from feeling profoundly rooted in his origins, while becoming a writer of the cosmic dimension. The hedge as a metaphor of marchigianità: because remaining on this side means being and feeling anchored to one's origins, while imagining, hearing and seeing everything that lives and pulses over the other side is to desire entry to the world, to become an active, not marginal, part of it.

The Guzzini family in the thirties: Enrico (centre), surrounded by children Celestina, Pierino, Silvio, Mariano, Maria, with relative spouses and children (note in particular Mariano's children: Giovanni, Raimondo, Virgilio, and Marcella).

From small business to industry

A family takes the offensive

By 1911 the new economic cycle (of 54 years) had almost reached its half-way point, therefore its growth peak. Enrico Guzzini was resuming his work in Recanati, after a few odd years spent in Argentina with his older brother and former wood pipe maker Giuseppe, who gave him the house and a third of a hectare of land in return for his work. Enrico was almost 50 years old, with five children, including *Pierino* (b. 1899) and *Mariano* (b. 1904); another son, *Silvio*, would arrive in 1913. The prospects were that his sons would eventually be able to assist company activities by the end of the decade, until the outbreak of the war[1].
The year 1911 witnessed the war in Libya under the fourth and second-last Giolitti cabinet. The following year, the right to vote was extended to males at least 30 years of age or who had completed their military service, even if property-less or illiterate: an electorate of eight and a half million, out of a population – with the contemporary borders, thus not counting Trento and Trieste – of 18.1 million men and 17.8 million women. A total of just under 36 million, just under three fifths of today's population.
The Marche Region had a population of 1,145,000 inhabitants, of which two thirds concentrated inland on plain lands (17.7% involved in industry, where work by women and children had been protected since 1886), just before two decades of emigration that would see the departure of over 200,000 from the Region. Data suggest that, since the time when Enrico was about 20, the "multiple" family had begun to lose importance, while the prevalence of births over deaths became increasingly significant and the *unmarried* were looked on with scorn, if not yet liable (as later, during the "demographic campaign", during the second half of the Ventennio) to pay a celibacy tax. Nonetheless, as will be seen below, the extended family in the Marche was still to express its greatest potential.

From horn to chemistry

Enrico and his sons crafted horn, a natural material, when artificial or synthetic alternatives were but in their initial stages of development. Unable to substitute natural materials, and in particular (above and beyond price considerations) those most prized by tradition, the new materials were subject to a certain diffidence (at the time not totally unjustified) in demand in relation to quality. This state of affairs would subsequently (on the eve of the Second World War) be reversed, with greater supply and demand for synthetic materials. Bakelite – introduced (so improperly) by Hyatt as a substitute for ivory in billiard balls – was at least half a century old, but its applications were more servile than noble: cutlery handles, then eventually for electrical appliances, thanks to its thermal insulation. It was however rarely used for decorative purposes.
The catalogues published by American department stores during the first 20 years of the century – such as by the largest mail-order chain Sears, Roebuck & Co, since 1880 (the catalogues available to us date from 1902) – featured not only traditional and noble (by today's standards) materials such as tortoiseshell and horn for the production of combs and other accessories, but also materials generally referred to by the term "imitation horn", or *hard rubber*, for use in items whose position in the catalogues themselves, their description and price were indicative of the lowly consideration accorded to them. Other materials, such as galalite and celluloid, would soon begin (especially after the First World

Cover of Sears & Roebuck mail-order catalogue, 1902.

No. 18R4011 Hard Rubber Dressing Comb. With fancy carved back; coarse and fine teeth. Length, 8 inches. Usually retailed at 25 cents. Price, each..14c
If by mail, postage extra, 4 cents.

Celluloid Amber Dressing Comb for 12c.
No. 18R4092 A Beautiful 7½-Inch Beaded Back Celluloid Amber Dressing Comb. Coarse and fine teeth. Druggists ask as high as 35 cents for these combs. Price, each. .(If by mail, postage extra, 2c)...12c

Imitation horn (plastic) combs featured in the 1902 Sears & Roebuck catalogue.

War and with the advent of *art deco*) to be used as substitutes – some even in objects deemed noble, such as costume jewellery and other goods with decorative purposes. Horn therefore remained one of the natural materials favoured for the production of items with traditionally aesthetic functions, with semiological requirements that accompanied their instrumental application.

Horn is a relatively inert material, and as such resistant to corrosion by acid. Its main qualities include hardness combined with flexibility, as well as variegated colouring, from white, through to brown and black (both features due to its fibrous anisotropy)[2].

For these reasons the material was used for highlighting (not unusual for other natural materials, such as wood) that, by limiting creative freedom, eventually refined skills and posed a challenge to the innovative tendencies of the producer; a specific ability developed in the precise selection of material in relation to applications and the achievement of special decorative effects. Given that the main methods for working horn involved cutting with special fine saws or moulding, the shape of finished products was inevitably classified (regardless of current fashions) into a relatively small number of *patterns*, the variations of which distinguished the creative skills of the artisan, who in turn responded to the growing demand for constant shapes and colours in production batches by traders. The significant cost for the material (mostly imported) led to the decision to direct (if allowed by the microfibrous quality of the batch) cuttings or rejects to 'lesser' items, smaller or less expensive: in short, a kind of distributional synergy.

The family grows

As said earlier, Enrico's activities – increasingly aided by Mariano and, following an apprenticeship with the Capodaglio, Silvio – had focused on the production of snuff boxes, in itself an important break away from the local comb-making industry.

In the meantime however, the family continued to cultivate their small plot of land, raise livestock and even to breed silkworms for the nearby mill. Mother Annunziata Frapiccini herself had performed humble chores to ensure an income, particularly during the long absence of her husband. In 1925, on Pierino's return from various vocational callings and the war, a family meeting decided to improve organisation and accelerate production, through the investment of 3,000 lire for the purchase of equipment able to triple the output of snuff boxes, as well as to diversify into the production of spoons and spatulas for table and cosmetic use, salad cutlery supplied to silversmiths for application of handles, long and short shoe horns.

Rectangular horn snuffbox by Azienda Guzzini (later Fratelli Guzzini), 1912.

As is often the case with new enterprises, high production levels soon filled the home, threatening the precarious family budget, despite orders arriving from as far away as Australia; at other times, however, even a single sale (such as to the *Pettinificio Marchigiano*) transformed a courageous business decision into success. This even without serious run-ins with the Capodaglio family, whose business fortunes were waning in the face of competition, in particular by cooperatives: the run-ins were in fact to come later, when the Capodaglio decided to include snuff boxes in their product portfolio, blatantly copying Guzzini types and models. The emergence of other competitors from further afield eventually triggered a price war. By 1929, the Guzzini had their first representative to tour the nation, though right on the eve of the Great Depression which, though arriving a few years later than in other countries, bankrupted *businesses* and made customers insolvent. These were also the years during which Mariano and Irene Guzzini's seven children were born: initially one every year as of 1927 – Giovanni, Raimondo, Virgilio – then more spaced apart, Marcella, Giuseppe, Giannunzio; then lastly Adolfo, born in the midst of the war. The Guzzini response to an increasingly 'perfect' competition and the general crisis consisted on the one hand in the exploration of new, more economically-viable sources (such as Thai buffalo) of horn in order to cut production costs, and on the other in the refinement of the model range, which also involved technical modifications. The third response to the crisis – which brought excess production to dangerous levels – was a trade agreement with the Capodaglio (who kept their core business of combs separate) and the opening of a sales office in the city (while production was increasingly in the hands of younger relatives, some forced to abandon their studies). But despite the general situation in 1935 – the enterprise had officially become a company the year before – the 60 square metres occupied by the factory (warehouses included) were becoming insufficient to house production equipment, the 20 or so workers plus raw material and product stockpiles. The decision was therefore made to extend to 150 m^2 by 1936 (becoming 200 m^2 by the end of the war).

The year 1938 – on the eve of the new world war, and while Enrico was gradually handing over the business to Pierino, Mariano and (when not serving in the army) Silvio – witnessed an opportunity that could hardly be described in any other way but crucial not only for the fledgling enterprise, but also for the history itself of materials and design (and not only in Italy): the adoption of Plexiglas sheeting for certain applications, particularly for salad cutlery. Produced by Vetrocoke of Porto Marghera on licence by Rohm & Haas of Darmstadt at a cost of 50 lire per kilogram, it was used to replace conventional material (at 2 or 3 lire per kilogram).

Salad cutlery by Azienda Guzzini (later Fratelli Guzzini), 1920s (Recanati, Fratelli Guzzini Museum).

The complex train of events that followed was instructive as regards resistance to change and to the processes of development and implementation of innovations (especially substitutive ones), even in sectors for which they were not originally addressed. While on the one hand we witness the prevalence of the so-called total cost of use[3] over the unitary cost in business calculations, on the other it gives an insight into one of the methods followed for introducing new materials and technologies that would characterise the invasion of plastomers in the period immediately following the war – and not only in small companies. Plexiglas was the commercial name given to a thermoplastic material – polymethyl methacrylate – produced as a possible substitute for glass, particularly in aeronautics applications, given its good transparency and resistance to impact, with a modest specific weight (the latter characteristic was what would ultimately favour the success of the plastomers). The idea of using Plexiglas for cutlery – an application for which it was not designed when initially produced – was one of the first cases of *technology-switching*[4], and indeed might even be seen as an example of *serendipity*[5].

As regards costs, it should be borne in mind that the innovation by the Schrumpeter-style businessman – i.e. one who is able to design new combinations of resources to operate profitably with competition, whether perfect or imperfect – assumes the total cost of use (deriving from the sum of all activities involved in the whole process of purchase-transformation-distribution) as the term of comparison in relation to previous ones, for which the unitary costs of a material or a service are but a part. In actual fact, high unitary costs are often more than not compensated by the reduction of other costs, including the famous price-cost (Cournot), by which an innovative product can be sold at a price significantly higher than that of conventional products.

The advent of Plexiglas for cutlery was one of these cases – one of the very first of the plastomers, in particular for domestic applications. The mechanism for the removal of the (understandable) resistance to change by all parties concerned (consumers included) was also paradigmatic, because it involved – typically – a business risk for three protagonists: the manufacturer of the material (who showed its applicability to new products by providing also rejects and cuttings deriving from the production of material to replace aeronautic glass sheets); the distributor (who ensured the absorption of at least one economic lot); and the transformer, caught between the first two, running a small risk of failure, though making investments for process research, production equipment renewal and the design of final products.

The collaboration that ensued clearly went to the advantage of all three: for the distributor (Richard-Ginori), this was one of the first cases of material diversification in an activity historically characterised by ceramic and porcelain, though commercially synergic given that it mainly concerned the catering business; for the Guzzini family, for whom it represented a solution that broadened the limited prospects offered by *labour-intensive* products (up to 30 processing stages for each piece) while solving the problem of quality fluctuations due to the anisotropy (mentioned earlier) of the horn and its origin (while potential applications would prove to be far more numerous); and for the manufacturer of the new material, for whom this would be the beginning of applicative marketing that future plastomer producers (at least in Italy) would be unable to implement completely even 40 years later[6].

The result for the Guzzinis following the first consignment was a remittance of 11,000 lire, for that era a remarkable sum (as the radio sang "Se potessi avere / mille lire al mese" "If only I could get / a thousand lire a month") that marked the beginning of an adventure that was to have an impact on the company and the future of plastic materials in Italy that was difficult to calculate. Not even a year had gone by before the Guzzinis called on the Capodaglios to shut down their joint commercial office, leaving them to their horn combs, while they prepared to plow – together with their usual portfolio of snuff boxes and ever-beloved horn cutlery, a production line to be discontinued much later – the new, vast plain lands stretching out before them.

Mould and press for manufacture of horn spoons by P.M.S. Guzzini (later Fratelli Guzzini), 1940s.

Following the initial acquisition of heat-forming machinery, the processing of Plexiglas posed new technical problems, which persisted until the introduction of a new method – designed independently by the Guzzinis themselves and other foreign groups – based on the use of compressed air. The new method, which prevented the material from deteriorating or losing its transparency, was implemented by Pierino, then subsequently applied to the forming of containers and objects of non-standard size, such as domes that the material supplier specifically allocated to applications in aeronautics.

A case of authentic 'reverse' *technology-switching*, associated with what would well after the war be acknowledged as one of the 'virtuous' achievements legitimately boasted by the Japanese. On the eve of the war the small company had begun to present – next to the traditional horn objects – a broad portfolio of

Moulding and finishing of classically-shaped container made from plastic material.

Plexiglas products designed for a range of different consumer categories: ladles, mugs and cups, cheese dishes, salad bowls and cruet stands, trays, items (including decorated ones) for children, marketed in many important cities in Italy. In fact the number of company employees once again began to grow, with the recall of the workers previously sacrificed by the crisis.

The war

The war was to separate the company first from the south of Italy, then subsequently from the north, as the front passed through the Marche Region to the Gothic line around June 1944. This was a dramatic, bloody period that sorely tested the vitality, the determination and the courage (when crossing the lines) of the Guzzini brothers – Silvio was serving in Albania, returning later to dedicate his time to horn product manufacturing – and even of their eldest children, in the effort to keep a foothold in the market, to glean what little money could be obtained and to tackle spiralling inflation (which was to reduce the value of the lira by 100), as well as to assure supplies of raw materials. Thus came to the surface the all-Italian and rustic ability of 'getting by' during difficult times, facing one day the Germans who wanted to confiscate material and destroy machinery[7], the next the Allies[8]. And of course, inflation.
The last load of Plexiglas (30 hundredweight, so much more precious then) before the advent of hyper-inflation was paid at the rate of 134 lira per kilo; the first purchase with the new prices was made at 3,600 lira per kilo, to eventually settle at 3,000 lira. Price setting for finished products was particularly difficult during the last year of the war, involving consideration of costs but also potential demand. Furthermore, the shortage of contacts and information sources were an encouragement to take opportunities that promised immediate benefits – even if they were in another field of activity. The important thing was having sufficient technical and commercial skills. This applied to various types of products for which there was demand at the time (religious *souvenirs* for the Loreto sanctuary, piano accordion keys, *gambling chips*), as well as to certain war-time products (thus providing a work opportunity to otherwise starving labourers and peasants). The survival of a company, a family and a microsociety was 'well worth a war'; researchers of war economies (and hyper-inflation) could indeed learn quite a lot from this type of experience.
The post-war recovery offered both the company and the family new prospects and gradual maturation of the in-house cultural climate. In the face of the prospects, the Guzzinis – without abandoning the first Italian supplier – immediately visited the Rohm & Haas plant in Darmstadt, purchasing relatively large quantities of Plexiglas at a more economical price. Soon after they opened direct relations with ICI in Great Britain to obtain the product Perspex, similar though available in a variety of pleasant hues. In this regard it should be recalled that production of methacrylates and other plastomers and materials in Italy, though licensed by foreign companies, was still a consequence of the home policy introduced by the regime in 1935 in response to the sanctions applied to Italy by the League of Nations following the attack on Ethiopia. Direct international contacts in the immediate post-war period were thus a new trend for individual businesses and for general trade prospects.
Ever since late 1950, the success of the portfolio-product policy implemented before the war – that witnessed the domination of the still dynamic horn pro-

duction by methacrylates – had led the family to decide on the construction (without any loan) of a new 500 m^2 factory, though still occupying the current site. The new facility, however, was designed not only to house more production equipment, warehouses and offices, but also apartments for the now numerous family units. The bond between family and enterprise, even in terms of logistics, remained steadfast, despite a certain degree of discord surrounding the renewal of office equipment that emerged with the third generation.

The post-war situation

Unlike the first war, the Second World War – the break-out of which virtually coincided with a new macro-cycle (1940-94) – had marked the development of new processing and product technologies, scientific and methodological studies, and above all the communion of scientific and technological activities. As of the post-war period, starting with the areas that can be defined as high-tech intensive in developed countries, it would no longer be possible to separate or distinguish between science and technology (and economy), nor to assign institutional and temporal priorities between them. The consequences for local, national and subsequently supra-national politics are easily imaginable.
During the war, two scientific-technological macrophenomena emerged with all their profound implications: the nuclear and missile technologies, with their respective significance for the energy and space industries. The subsequent race to obtain the relative scientific *know-how*, necessary to develop applications, vital as diplomatic deterrents and for image, led countries that prior to the war had allocated greater resources to the development of processes and quantity (such as the USSR) to enter the typological and qualitative competition of 'products', even if mainly for collective use.
At the macro level in the 1940-96 cycle, on the one hand technological innovations increasingly followed the results of scientific research conducted in assigned facilities; on the other, science and technology took on an increasingly important role in international competition. An impressive quantity of achievements would consequently appear: new materials, equipment, methods and forms of organisation of both research and production[9]. Due to their dimensions and variety of forms, these achievements can only be mentioned briefly, attributed to the usual USA (and later Japan, which boasted a reputable tradition of scientific and technical research even before the war[10]) and European countries, such as Great Britain and France (and Sweden). As regards Germany, research was limited by the conditions of surrender, the destruction caused by the war and the diaspora of scientists (not only nuclear) during the last quarter of the previous cycle[11].
In Italy, the need to rapidly reconstruct production facilities, the general economic state and the organisational chaos that followed the war distracted attention away from the demands of research, compounded by the lack of comprehension by the political establishment[12]. But after the initial state of confusion, after 1951 internal demand began to show particularly high rates of growth: between 1951 and 1958, an average of 5.5% per year, a recovery led mainly by the building industry, public works and agriculture. Between 1958 and 1963 however, growth surged to 6.3%, with a doubling of industrial production, while exports increased at a rate of 14.5% per year. In 1951, 18,500 refrigerators were produced in Italy; by 1957, this number had grown to 370,000, a prelude to the

veritable explosion in production culminating in 1967 with 3,200,000 units. In 1957, 151,000 typewriters (standard size, practically the equivalent model to the *Lexicon* Olivetti) were produced, the output reaching 652,000 by 1961. Overall production of plastic materials during the same period expanded by 15 times. Nonetheless, right up to half-cycle (1967), technologies, though having enormous potential for civilian applications (such as the controversial nuclear power plants), did not lead to any trickle-down effect from the macro to the micro, while technological exchanges between sectors that would eventually influence the technological policies of states and supranational bodies (e.g. the EEC) were sluggish. Furthermore, right through to the second half of the cycle, society – engaged in reconstruction, with priorities that varied from one state to another – remain 'detached' from technological evolution, particularly macro scientific developments. This was partly because scientific language – in parallel to that of the arts and the critique – had become overly sophisticated without any popular press (and not just in the vernacular) to explain it. The same applied to technological evolution because – save in a few sectors (e.g. in Italy, the automobile, motorscooter and typewriter industries) – the impression was that of a trend towards the perfecting of already-familiar goods in a context of incomes so depressed that their initial distribution was incredibly slow.
During the first quarter of the cycle (until 1953-54), society in general – but particularly in Italy – merely resumed the value hierarchies and 'daily practices' of the pre-war era with a priority view to achieving reconstruction as rapidly as possible – where the term reconstruction (not the case of the Guzzini brothers) was used in the literal sense, i.e. restoration of the past, now remote, of the prewar period. In Italy this even meant the pre-Ventennio (the twenty years of Fascist rule), given that, in general, the political establishment insisted in its rhetorical appeals to Humanism, as if material culture was antithetic to it. And yet, despite the lagging educational system and the shortage of information regarding the interpretation of the revolutionary meaning of technology, that period can well be considered the one that heralded in a new basic model for development beyond the half-way mark of the cycle.
The widespread detachment/disinformation on the state of science and technology led society itself to believe that it was somehow ahead of technology – a conviction that would have a severe impact in the social crisis half way through the cycle (1968). The generational changeover – meaning the change in attitude and the rise of young ideas (including economic ones) – took place slowly, at least in Italy. Even the design of new products witnessed the intellectual prevalence of figures young before the war: in large corporations, the same businessmen, managers and design engineers came again to the fore. In small and medium-sized companies, the parallel was far less obvious, the generational changeover more rapid: designers and engineers considered young before the war (and those who had entered the profession with the greatest esteem for the people considered the 'maestri') came increasingly into contact with new entrepreneurs. This in turn led to the phenomenon (not only in Italy but also the United States[13]) of innovation led by small and medium-sized companies, connected (or not) to larger companies. The phenomenon was most deeply felt in the furniture and, in general, in the furnishings trade (for the most part made up of small businesses), especially at the turn of the second quarter of the cycle – that now legendary year of 1954, that marked the launch of the industrial design industry[14].

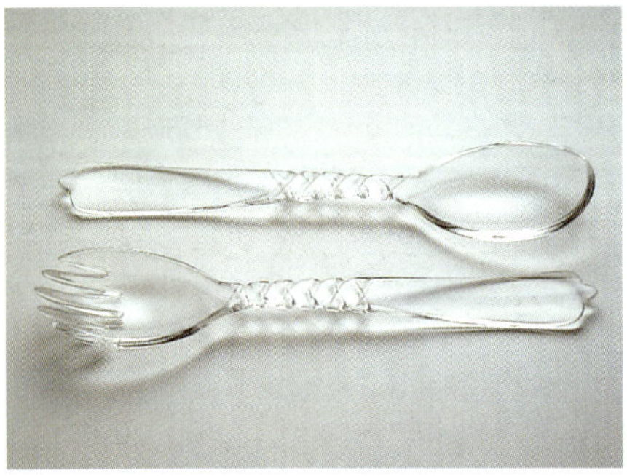

Top to bottom: High round bowl (1941).
Salad cutlery with twist (1938).
Cup and saucer, Disney tea party set (1950).
Crafted by P.M.S. Guzzini (later Fratelli Guzzini).

The family evolves: the Guzzinis

The term "family" has two meanings: a synchronic one, that denotes effective bonds, of solidarity and even of antithesis, among the few or many living components; and a diachronic one that is based on continuity, a common culture, attitudes that evolve from one generation to the next. In both senses, the Guzzinis have always been, more than anything else, a broad and complex family, with explicitly or implicitly driving personalities – that are not always easy to identify. It is for this reason that the family emerges as a value, a *Gestalt* therapy, in which the sum is less than the individual components combined.
In the years leading up to the war and even during it, the brothers Pierino and Mariano were the driving force behind the family. Much has been written about Pierino, his dynamic activities – technical, commercial, administrative – that depict an enthusiastic yet practical person, known for his business-like ways that gave priority to operativity. Mariano however left no written testimony of his family and company experiences; if the author is not mistaken, to do so would have been out of his more introverted and introspective character. Nonetheless, what is known is that Mariano was certainly involved in production and trade activities, often acting as moderator and coordinator in his cautious, laconic manner, though with managerial courage and open business outlook. Though hard to find two more different people, it is easy to imagine how, when working together, their personalities complemented one another. But in the face of the enormous internal and external changes of the times, it seems logical that the approach of the latter, Mariano, would prevail; as much transpires from the words of many, and not only from his sons. On his return from the front, Silvio also made his contribution to the decisions tackled by his older brothers to ensure the development of the family business.
The following family members began to work at the factory (mostly at an early age) with labour duties: Giovanni (b. 1927) as of 1941 at the age of 13, to fill in for workers drafted into the army, then increasingly dedicating his attention to technical design and production, with a particular eye to market demands; and towards the end of the war, Virgilio (b. 1929), whose creative skills and instinctive curiosity enabled him to transform what is defined today as *technology-push* into *market-pull*, and vice versa. His most notable achievements in collaboration with Pierino and Clemente included the first casting furnace for the creation of the monomer, which made the use of plastic sheet material no longer indispensable. Personal accounts indicate that the preferred procedure involved following up a manual apprenticeship with the assignment of various duties that combined personal skills with other activities, thus ensuring all-round business training (that did not however take them to activities outside the company itself, as will be seen below). Raimondo (b. 1928) was in fact the only one to immediately pursue the corporate career for which he felt most talented; in 1950 he returned from a working stint under a Neapolitan representative to acquire commercial skills, which he was later to use with keen strategic sense. In 1954, it was the turn for Giuseppe (b. 1935), fresh from accounting school, and Giannunzio (b. 1938). Giuseppe had the responsibility of administrative and financial management of the company. To this end he introduced the first mechanised administrative processes, while producing the tools necessary for monthly management control and the drawing up of the annual financial statement. His conciliatory nature also gave him the role of coordinator between the various families. With their youngest brother Adolfo (b. 1941), they entered the com-

Mariano's family at the entrance to the Fratelli Guzzini factory, 1960s.

pany during the second half of the fifties, from the exterior as it were, on the appearance of a new and separate activity undertaken by all of Mariano's children: a production that involved a much more sophisticated form of injection moulding than had originally been adopted to compensate the problems encountered with two-colour processes, to which Raimondo had dedicated his renowned market sense. Witness the turnabout in the logic of marketing strategy; in fact Raimondo was one of the first to realise that assigning this task to a prestigious department store would give far greater market penetration to new products. Obviously, the store could not be any but la Rinascente, recent inventor of the Compasso d'Oro award for design excellence and organiser of international exhibitions (Japan, Spain, England, the USA, Latin America) that were to make its fame.

In actual fact, it was as early as 1954 that the Guzzini brothers acquired and initiated production of the first manned thermoplastics moulder, a British Windsor machine that gradually underwent so many modifications (including original ones by Alberto) that it was eventually transformed, together with others subsequently purchased, into a totally automatic unit. One might well ask why this interest in automation at a time when manpower was abundant and relatively cheap?

The answer gives food for thought, especially for its historical and business implications. First of all, there was, as always, a spontaneous thrust in that direction: innovations are also due to the sheer enthusiasm of the company employees – even though certain skills would not have been encouraged if there had been no documented aversion to seeing operators 'serving' production equipment, virtually immobile. Another strategic level – observed by others (especially the younger members of the family) – was the awareness of the competition threat by so-called 'closet' industries characterised by uncertain salaries and low running costs, conditions later to be defined as almost "submerged" – a typically Italian phenomenon, soon to be widely adopted, and even enhanced, by the New Industrial Countries of Asia. One might object that the product policy, later characterised by a greater interest in diversity than in actual price, should not have generated any fear of such general *price-oriented* competition. But this would ignore the fact that, while some maintained that the business logic was based on savings (while perhaps accepting time-related – though overtime – costs for design and testing[15]), others argued that the production of originally 'designed' items would demand specialised processes at costs no higher than those incurred by outsourcing. Even though both product and process innovations (i.e. perfect and imperfect competition matched and integrated) have been constant features of the Guzzini history, a clash was inevitably brewing between the different business approaches and views towards the end of the so-called 'Italian economic miracle'.

Sauce boat with two-colour spoon (1954), one of the first products to be made with the new technological process patented by Fratelli Guzzini.

The year 1954, at the conclusion of the initial stage of the current macro-cycle, was thus especially important for the Guzzinis, who were just beginning experiments on the polymerisation of methacrylate from acrylic monomer by *cracking*. The results would allow the cost-effective production of sheets with different thickness and characteristics as of 1958, as well as of two-colour[16], a Guzzini original. Indeed it might be said that 1955 was the year that commercial responsibilities were fully entrusted to Raimondo, given that he accompanied his father and his uncle during the important visit to Germany (naturally immediately to Darmstadt), while in 1962 he brought in Luigi Massoni, then just over 30 years old, to develop company design.

While the third generation was emerging, in 1958 the initial launch stage – which for the Guzzinis had lasted an entire macro-cycle, marked by direct experience of an economic crisis and a devastating war – thus drew to a close, followed by the opening of a new phase of great economic expansion, peaking in the years of the "hot autumn", 1968 and 1969.

It was in this context of current and prospective development that in 1958 the family was called to make some crucial decisions; a testing time for family unity, with the emergence of two different visions of a modern world that demanded adequate solutions and proposals.

A company document succinctly describes the dissenting intentions and wills of the various segments of the family: those who felt protected by an attitude considered protective of others, and those oriented towards a bolder, more modern entrepreneurial approach.

Even though the bone of contention was the development of a business successful thanks to the contribution of all, it would be overly simple to dwell on the more obvious aspects of the choices made, such as whether to adopt new technologies requiring more investment than employees[17]. It was rather a matter of deciding whether to continue a strategy founded on industriousness and aided by techniques (even new ones) in slowly-evolving markets, or to embrace a line of development in view of rapidly changing demand for quality and quantity, using the new techniques as an instrument of its own acceleration, i.e. take active part in change. On encountering the refusal of other family members,

Domestic luminaires designed by Giancarlo Capici (1964) and manufactured by Harvey.

Mariano's sons proposed an intermediate solution which, while apparently a compromise, actually reiterated the need to broaden the 'vision' beyond the company's dependence on the household goods' market[18]. But even this idea was rejected.

It was thus that a small company, Harvey Creazioni, was bought into by Raimondo, Giovanni and Virgilio separately from the other two branches of the family, then purchased outright. Giuseppe and Giannunzio also entered management, as did Adolfo after completion of his studies. The company produced methacrylate lamps and enamelled copper items. The workshop was set up in Giovanni's cellar, where the young ones "worked from six, six thirty until midnight". The venture was successful because it led the family to reconsider, even though by now the younger members were convinced that it was insufficient to change what was already in progress. So to preclude future disagreements among members (around twenty of them) of the third generation, or worse, damaging competition, it was wisely decided that any family member was free to open new businesses, provided that they did not compete with pre-existing enterprises. The result was that, while Mariano's sons continued to develop their business without actually abandoning Fratelli Guzzini (in which everyone had a share), Pierino's sons opened Acrilux, for the sole production of acrylic luminaires.

Notes

1. The information on Enrico Guzzini and the early years are partly taken from a long memorial penned by his son Pierino in 1976 and kindly made available by the Guzzini family.

2. The material is derived from the exterior of the horny casing of the frontal eminence of the male of certain ruminants, such as cattle and sheep, mainly from Nigeria, South Africa, Argentina (except in the case of the reindeer, the female of which also grows antlers). Anisotropy – i.e. the inconsistent nature of materials in all segments (isotropy) – is a widespread characteristic that slowed and eventually prevented the industrialisation of goods made from materials of plant and animal origin, while favouring (amidst other things) the development of new artificial and synthetic materials, such as resins and thermoplastics.

3. See A. Morello, on "Marketing", Milan, 1980.

4. "*Technology-switching*" denotes the use of a method or material (which as a rule involves a specific processing technology) in sectors other than those for which it was originally designed. *Technology-switching* is to be considered one of the most important phenomena in the second and third stages of the industrial revolution, having generated opportunities for improving production processes while extending the use of technological product advances, as well as increasing productivity and profitability of scientific and technological research.

5. "*Serendipity*" derives from the Singhalese "blessed moment of good fortune", and is used to describe the phenomenon of finding something useful not sought for. The importance of *serendipity* is documented by countless innovations. It is considered an intrinsic element of the Schumpeter-type entrepreneurial skills.

6. During the years 1975-85, the author still encountered difficulties in having integrated and widespread marketing solutions adopted (in terms of application area rather than plastomer) by material manufacturers.

7. Of the numerous events in Italy of that period, it is worth noting the intent (announced by the Wermacht) to confiscate the larger Plexiglas sheets, to be used for military purposes. When the SS actually arrived, the quantity available had been reduced as much as possible to avoid accusations of opportunistic conduct. Following intelligence reports, the machinery was carefully lubricated, packed and buried in a field of maize, where it remained until late August 1944, when electric power facilities were restored to operating efficiency, albeit irregular.

8. Between 1944 and the end of the war, the market had been reduced to the South; material supplies had in fact been held up by the Allies on the Gothic line, while arrivals from the Rohm & Haas Philadelphia facility had been interrupted due to the lack of import licences.

9. The main inventions and innovations of the post-war era include: the ball-point pen (Great Britain 1944, then Làszlo Biro, 1945); the microwave oven (Raytheon, USA, 1945); the instant camera (Polaroid, Edwin Land, 1943-47); the telephone answering machine (Buhrle, Oertlikon, CH, 1949); disposable cellulose tampons (USA, 1949, then Procter and Gamble 1961); trunk dialling (Leslie Denning, Bell Telephone, 1951, then AT&T, 1970); the geodesic dome (Buckminster Fuller, 1951); cable television (USA, 1951); tranquillizers (reserpine, 1952); synthetic diamonds (General Electric, 1955); the transistor (first application, Bell, 1951; first radio 1955); Liquid paper (Bette Nesmith Graham, typist, USA, 1956); Velcro (George de Mistral, USA, 1957); the skateboard (B. and M. Richards, 1958); the pacemaker (A. Senning, Sweden, 1959); the audiocassette (1929, produced by RCA 1958); the fax (AT&T, 1964); aseptic packaging (Tetrapack, Sweden, 1964); shop anti-theft device (A. J. Misnay, USA, 1966);

liquid crystal display (Hoffmann-La Roche, Switzerland, 1950-51); pocket calculator (Canon, Japan, 1970); axial tomography (G. Houndsfield, Great Britain, 1972); the videogame (USA, 1972); the plastic soda bottle (1973); the videocassette (Philips, Netherlands, 1974); the bar code (Ohio, USA, 1974); in-vitro fertilization (Cambridge, Great Britain, 1978); the Post-It (3M, USA, 1979); the Walkman (Sony, Japan, 1979); the compact disk (Philips, Netherlands, 1980); genetic fingerprinting (Colin Pitchford, Great Britain, 1987). Information Technology: the "Great Brass Brain" (USA, 1910); Differential Analyzer (USA, 1935), Colossus (FRG, 1948); EDVAC (USA, 1950); COBOL (language, Grace Brewster Murray, 1950); UNIVAC (USA, 1951); Whirlwind (USA, 1951); Internet (network, 1973); BASIC (language, 1975); Cray 1 Supercomputer (USA, 1976); Wordstar (USA, 1979); Dbase (USA, 1980); MS-DOS (USA, 1980); Lotus (USA, 1982); the Desktop Laser Printer (Xerox, USA, 1984); Windows (USA, 1985); Virtual Reality (1985). The stages in the development of the desktop computer (Apple, Commodore, IBM, Olivetti, etc.) are not listed as they are already widely known.

10. This fact was already obvious from the naval fleet catalogues of the *Lega Navale Italiana* dating from the late thirties, in which the very form of Japanese ships featured an unprecedented technological and organisational design. As regards science, reference must inevitably be made to Nobel prize winner Hideki Yukawa, who had theorised the existence of the meson (later discovered by Anderson in 1937) at the same time as the Fermi school in Italy (1930-38).

11. In occupied Germany, the Allies were mainly concerned with dismantling the production and research plants set up by the Third Reich, such as IG Farben, which was broken down into Bayer, Hoechst, BASF and other minor companies. During his studies on plastomers for ANIC (1978-82), the author formulated the hypothesis (still apparently valid) of the formation – amidst the three main manufacturers – of an oligopoly devised to protect price levels through the carefully-metered supply of quotes by each in common polymers: a classic response by a corporate strategy oriented towards (and by) the market.

12. As seen by the shortage of public initiatives to assist a rapid resumption of university laboratory activities, at least to the level attained before the war. This general insensitivity regrettably did not exclude Luigi Einaudi, who in 1947 turned down a request for funds to scientific research; indeed the lack of commitment to research by the CNR eventually led industry to look elsewhere.
It should also be recalled that in 1947-48, the CNR received investment funding of only one thousandth of national income, while allocations in the USA were around 1%. Of note the protests lodged by Luigi Morandi (Montecatini) and Vittorio Valletta (FIAT) with the ministero per la *Costituente* (1946). Cf. R. Maiocchi, *op. cit.*, p. 960 ff.

13. As indicated by a study conducted at Harvard (USA) relating to the sources of innovation for small, medium and large-sized enterprises (1970).

14. As regards the events witnessed in the furniture industry in Italy, see *Il Salone del Mobile 1961-1986. Le diverse età dei linguaggi*, 1986.

15. The Italian jocular remark "Spare no expense, as long as you save money" is symptomatic of a particular approach to business. Another one might be quoted, though with a slightly different meaning: "Spare no expense, as long as you make money". The author leaves the choice up to the reader.

16. In the early fifties, Gaetano Clementi, a reputable chemist who owned an acrylic sheet plant in San Giovanni Valdarno, by fortunate chance (a brochure found in a waste bin) came into contact with the Guzzinis. From there an important partnership was born that, first with cousin Alberto, saw Virgilio Guzzini work for 13 years in a production unit constituted *ad hoc*: Polivar of Pomezia, to which the Guzzinis made a fundamental contribution, as shown by the words of Clementi himself.

17. "Towards the end of the fifties, Mariano's sons decided to set up a new venture. [...] Raimondo had suggested that new market demands needed to be satisfied (e.g. acrylic sheet lamps), that technological strides needed to be taken towards injection moulding. This meant a transition from small-scale business to industry, from a labour-intense system with little capital to something completely different, requiring large amounts of capital to build plants and moulds, purchase machinery, build up stocks, with lesser manpower." (company document quoted previously).

18. "In the meantime [Mariano's sons] proposed a diversification in the production of household products, the manufacture of interior light fittings, using methacrylate combined with other materials, such as wood and metal." (company document quoted previously).

The culture of industry

1949: brand new Vespas, designed by Corradino D'Ascanio, leaving the Piaggio factory.

Project and design in the immediate post-war period

The *motorscooter* was without a doubt the first widespread innovation of the post-war period; a mass phenomenon destined to change the Italian landscape as a reply to the demand for individual mobility at a low private investment cost. The history of the *Vespa* designed by Corradino D'Ascanio (1948, immediately followed by Cesare Pallavicino's *Lambretta*) and its origins, opportunities and social repercussions are well known and require no further comment. As regards however the 'design', it was emblematic of the conversion of industry to civilian production (Piaggio), marking the difference between the formal configuration of the Vespa (whose name, "wasp" in Italian, significantly recalled its shape) and the *structure* of the Lambretta. The choice of one model or the other was often determined by social considerations: the slightly more expensive Lambretta was preferred for example by architects in relation to the rationalist domination of the younger generation.

The *user-identification* process, a phenomenon only recently emerging, thus continued and became more pronounced, indeed destined to reach unprecedented proportions in a context of freedom – but a *freedom that was ambiguous* because, though seemingly a principle assured by competition, it was actually oriented by price differences, the relative efficiency of distribution and the pressure of advertising and promotion[1].

The same 1948 saw the release of the *Lexicon* 80, followed two years later by the Olivetti *Lettera 22* designed by Marcello Nizzoli. The Lexicon 80 was a technical object that radically modified conventional perception: the machine was merely the intestine – though accessible through an opening that allowed movement of the type bar and became, by its shape, a virtual 'handle' for lifting off the panel for maintenance – of a unitary body characterised by its sculpted, calibrated contours and technically innovative housing (in die-cast aluminium) that emphasised its three-dimensional presence, not to mention its organic evocations (e.g. the carriage lever which had become a 'technological bone') and scoring at the rear, particularly when the machine is straightened (mainly when used by reporters) on the table. The *Lettera 22*, while having similar features and housing materials, had a radically different appearance: seen when standing opposite the work desk (or from the floor), it became a graphic page, flat and contained in an ideal grid, which emphasised its compactness. Every feature was simple, with a recalibrated type bar opening and essential carriage handle, which like in an insect became a cartilage. The light colour of both models courageously countered the usual black, while their very names – thanks to the remarkable lexical skills of Franco Fortini – were eruditely evocative. The advertising campaign – which had already created a sensation with the famous rose in the ink well, symbolising the obsolescence of writing by hand – that straddled the war period would become, under Xanti Schavinsky, Leo Lionni and later Giovanni Pintori, one of the most singular forms of artistic expression of the day.

The slogans themselves marked an era; recall the one devised by Fortini for the *Lettera 22*: "Leggera come una sillaba, completa come una frase" ("As light as a syllable, as complete as a sentence"). The impact of these two machines on design (and not only in Italy) was enormous – not so much for their distribution (much more limited than the *Vespa*) as for the change in shape and truly innovative design procedures that went beyond sectorial boundaries. While inside the mechanical writing industry they constituted a radical (and selective) transformation in appearance and function, outside the industry they embod-

Olivetti Summa 15 adding machine by Marcello Nizzoli in a poster designed by Giovanni Pintori (1949).

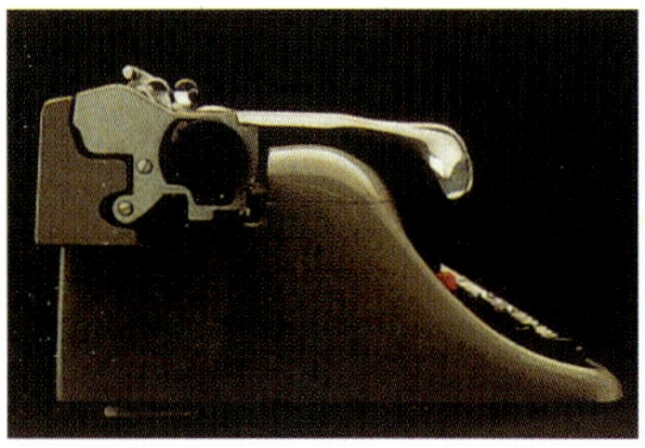

Lexicon (1948) and *Lettera* 22 (1950) typewriters designed by Marcello Nizzoli for Olivetti (Arch. Olivetti).

ied a completely new approach: housing that was not limited to covering and concealing, but also revealed and gave order to the configuration of the object; instead of formalising it (alone), they structured the relationship between the object and the user[2]. Furthermore, besides the product/advertising synergy mentioned earlier, Olivetti advocated an overall principle that captured the interest and cooperation of at least two generations of intellectuals (architects, sociologists, economists, philosophers, academics[3]) who, directly or indirectly, from Ivrea or elsewhere, looked to Adriano (1901-1960) as a champion – recalling both Robert Owen and Walter Rathenau – of that form of enterprise which, on the border between industry and politics, develops through a conscious cultural design. The issue was soon taken up again by Nizzoli with his two Necchi sewing machines, while, at Borletti, Marco Zanuso (b. 1916) continued – in the wake of his initial experiences in the application of new materials, such as foam *rubber* in the *Lady* armchair – his pursuit to simultaneously reveal function and structure by the *configuration* of a systematic, 50-year-long attempt at the conversion and highlighting of technical-industrial products and processes, in the search for archetypes. The automobile had its own standard bearers who, while proposing stylistic solutions obviously addressing the upper reaches of the market, explored configurations designed to imbue the automobile with an identity that was not only formal; witness the timeless Dusio Cisitalia (1947) by Battista Farina (aka Pinin)[4].

It hardly requires recalling that automobiles, *motorscooters*, typewriters, sewing machines and home appliances were goods with a structural and functional complexity far greater than that of furniture or household objects. The ambition of the latter was of course far more limited, given their traditional bond with home chores, at most reinterpreted, for example, by Gino Colombini (buckets, colanders, trays and apple graters) for Kartell. This company, incorporated in 1949, became a veritable symbol of the post-war era, even anticipating business practices (such as outsourcing) that would become commonplace only much later.

Giulio Castelli, a former pupil of Giulio Natta at the Politecnico di Milano, had in parallel to Ziegler discovered *isotactic* polymers (subsequently leading to polypropylene and the Nobel prize), had an idea: to produce only what could best 'serve' in plastic, but unlike others, in whatever plastic was suitable[5]. By the way, Castelli's personal account of his friendship and mutual esteem for Raimondo Guzzini reveals – possibly for the first time – a bond, but also a strategic antithesis in the use of materials, aspects that have always escaped the attention of contemporary design historians.

In any case, the materials in contemporary production – in particular the new synthetics that began to appear in great quantities – certainly played an important role in driving development, as seen in the opening of basic Italian production facilities, operating first on licence, then later with their own specific characteristics, in large chemical industries such as Montecatini and ANIC (ENI) as much as in minor ones (e.g. Mazzucchelli). In the meantime material processing equipment (often of Italian design and production) for injection and *blow-moulding*[6] was increasingly widespread.

The new synthetic fibres – polyamide types (Nylon), then polyester (Terital) and acrylic (Orlon, Dacron) – soon took their place beside other artificial materials of cellulose origin (such as Rayon) already widely used before the war (e.g. at SNIA Viscosa), partly due to the thrust of the autarchy that had followed in the wake of the sanctions applied by the League of Nations in 1935 in response to

The Italian household in the early sixties.
Above: articles made from Moplen (polypropylene produced by Montecatini) displayed in front of a shop (photo by Cesare Colombo).
Below: photograph advertising a Candy washing machine (from the "Necessario indispensabile" show, Milan, Rotonda della via Besana, November 1991 – January 1992).

the war in Ethiopia. The major Italian material producers nonetheless proved to be somewhat short-sighted (with due exceptions, such as Natta's success pursued by Montecatini) and slow-footed when it came to commitment to research. Such a state of affairs was to culminate in an international policy typical of the *price-taker*, i.e. importing of 'noble' or more innovative materials at a high price, coupled with exporting of *commodities* at a lower price, even when the balance of trade in the industry was in Italy's favour. That lively era nevertheless bore witness to the birth of the *modern Italian industrial project*; what in various degrees of complexity would make the world look at Italy, a country not particularly endowed with technology, as a source of intense development of industrial innovation for finished products. The economists, who have never been particularly clear as regards the role to be assigned to the configuration of goods (and probably, at least initially, to technological innovation[7]), from that moment onwards began to acknowledge that *imperfect competition* (i.e. not mainly based on prices) was becoming an increasingly important economic practice. In the seventies and eighties, they would finally assert their growing interest in their own concept of design, though not always lucid and often with journalistic intent. Moving from a situation of neglect, the *material culture* in fact ended up playing (at least in lip service) a central, though general, role in the declarations of economists, corporate consultants and even politicians. Since the late seventies, not one conference on design has been held without someone feeling the need to express their encouragement to the business community (which itself has taught others of the importance of design), then later perhaps confusing design with the concept of *made in Italy* (or even the *Mediterranean diet*).

The home appliances which – beginning in Italy with Borghi (Ignis refrigerators) and Fumagalli (Candy washing machines) since 1946 (followed up by San Giorgio under IRI and Rex by Zanussi) – constituted with others (the washing machine presented in Turin in 1950 by Riccardo Bertolino was commercially unsuccessful) the first nucleus of an industrial sector destined to export as much as two thirds of product output. Authentic serial production lines were the result, such as the Ignis plant in Cassinetta di Biandronno (Varese) in (once again) 1954[8]. In the meantime, new radios and gramophones were also being produced.

Even before the outbreak of war, the remarkable Phonola by Luigi Caccia Dominioni (b. 1913) and the Castiglioni brothers Livio (1911-1979) and Piergiacomo (1913-1968) had virtually revolutionised radio design: a metal or plastic parallelepiped housing connected to a loudspeaker set to 45° making it suitable for installation on a bench or hung on a wall, while tuning was for the first time ever controlled by keys. The concept was clearly a precursor to Nizzoli's typewriter designs. Only at the beginning of the second quarter of the new cycle (only by the braking effect caused by the choice between PAL and SECAM systems and the costs for the relative sound system) did television finally appear (once again in 1954), even though testing by EIAR (later to become RAI) had already begun in Turin in the late twenties.

After the first experiment by Piero Bottoni (1903-1973) in the QT8 Experimental District of Milan (1946-53) and the Triennali between 1947 and 1951, the 10th Triennale of 1954[9] organised by Marco Zanuso and Carlo De Carli duly dedicated new attention to international production. There can be no doubt that the event – by its innovative character as much as the way in which it was held – was one, if not the, most important of the post-war years. At the top of the staircase of honour (1933) by Giovanni Muzio (1893-1982) a cloud of solid white

First logo for the la Rinascente Compasso d'Oro award, designed in 1954 by Albe Steiner in collaboration with the company advertising department.

gauze chains by Piergiacomo and Achille (b. 1918) Castiglioni announced the first expo to also take a look at Italian design in the general context of the time. The convention that followed – chaired by the philosopher Enzo Paci (1911-1976) – might even be considered the manifesto of Italian design, though not actually in written form, given that the proceedings were never completely published[10]. Nonetheless, besides the first global plan for economic development (known as the Piano Vanoni) and the inauguration of national television service, the year 1954 also witnessed other key events in Italian industrial design, including the birth of the magazine "Stile Industria" which – taken from Gio Ponti's "Domus", the influence of which on the design culture was, together with "Casabella", virtually incalculable – until 1963 accompanied, recorded and even preceded the development of a new design consciousness in Italian culture, thanks in particular to its editor-in-chief Alberto Rosselli (1921-1976). Not to mention the first ever la *Rinascente Compasso d'Oro prize* – awarded for the first time at the 10th Triennale – to which only the leading names in the industry contributed (also in the way of juries, as well as winning designers and enterprises)[11]. The department store la Rinascente actually developed the Compasso d'Oro through the works of Cesare Brustio (b. 1910), in the wake of great attention to the issues of material culture which, having already appeared before the war by the work of Gio Ponti and subsequently Carlo Pagani, had intensified in the years immediately after the war – not only in the activities designed to boost sales, but also in the policy governing product assortments (following the setting up of a Directorate for the design of goods produced by suppliers), making it one of a select few of international distribution facilities to develop a real policy of goods innovation. Of the main European Department Stores to achieve a similar level during the fifties and sixties, reference can be made to Nordiska Kompaniet (Stockholm), Illum Bolighus in Copenhagen, and other minor stores. Only many years later would the Amat brothers – who declared their inspiration by la Rinascente – come up with the remarkable and ever successful Barcelona chain Vinçon[12].
The Italian Association for Industrial Design (ADI) was formally inaugurated in 1956, by nine designer-architects (Carlo De Carli, Ignazio Gardella, Vico Magistretti, Angelo Mangiarotti, Bruno Munari, Marcello Nizzoli, Enrico Peressutti, Alberto Rosselli and Albe Steiner), two businessmen (Giulio Castelli, Kartell; A. Pellizzari, of the enterprise of the same name in Arzignano) and a critic (Gillo Dorfles). Unlike other designer bodies, the ADI was made available as a venue of cultural debate amidst all protagonists of industrial design. A similar venture was launched in London in 1957 with the constitution of the International Council of Societies of Industrial Design (ICSID), which held its first congress in Stockholm two years later.
During its early years, ADI was mainly sustained (besides the two already mentioned) by the companies Olivetti, la Rinascente and Pirelli, while others such as FIAT, Pininfarina, Zanussi and Brion-Vega would arrive some time later[13]. ADI first formally received the Compasso d'Oro prize – awarded by la Rinascente – in 1959 (though in actual fact in 1962), first for *product design*, though only later for *industrial design* excellence.

Winds of war

In 1959, in the midst of the second quarter of the cycle, the panorama of Italian design, which was prevalently characterised by an evolution still (oddly) connected with the rationalist movement of the pre-war years and the lessons of the latest *Bauhaus*, was poised for change by the appearance of certain products and designers (mainly architects) who expressed a formal predilection increasingly oriented away from the geometries that defined the space occupied by furnishings as the organiser or prompter of human activity. The attention dedicated to space by Albini, BBPR ([Banfi], Belgiojoso, Peressutti, Rogers), Munari, Rosselli, Carlo Scarpa, Valle, Viganò and Zanuso (and with the chiavarina *Superlight* chair, by Gio Ponti, certainly not a Rationalist by any stretch of the imagination) was opposed by a new *focus of the design object in itself*, its specific shapes: thus, from the diastole of a semiology aimed at underscoring *syntactic*, design was marked by a new *semantic* systole, while the new furniture (and other) proposals returned to the centre of design interest, in the attempt to recover interpretative (and formal) freedom for productive processes that – to be recalled for the sake of honesty – proved only potentially industrial, given the artisan-type structure of a large proportion of furniture and of very little of mass produced items. It was no accident that this approach encountered immediate opposition by authentically industrial (or rather mass-producing) enterprises, such as in sectors reliant on plastic materials, at least prior to the introduction of new 'paddings' (mainly foam rubber).

Superleggera chair by Gio Ponti (1956, prod. Cassina).

On a more conceptual level, the first manifestations of what was later defined as the *neo-liberty style* was significantly led by the Lombard Giotto Stoppino, the Piedmontese Vittorio Gregotti and Lodovico Meneghetti (*Cavour* armchair, 1960), who the same year invited the Turinese Roberto Gabetti and Aimaro Isola (*Casa Gabetti* beside table, 1959) to the *New Design exhibition for Italian furniture*. The original 'piemontesità', the Piedmontese character, of that experience cannot but recall the contradictory and charismatic figure of that bourgeois viveur, so endowed with imagination and formal sensitivities, that was Carlo Mollino (1906-1973), lecturer at the faculty of architecture of the Turin Politecnico between 1953 and 1970. Whether or not everything that was (and still is) defined as *neo-liberty* was actually connected to Liberty, the events of the time were an important yet problematic break; not only because of the new focus on semantics, but also because for the first time since the war style was once again being discussed, in a certain sense in opposition to the broader meaning of the term *language*. Amidst other names, Achille and Piergiacomo Castiglioni themselves seemed drawn by the vague of the moment in the *Sanluca* armchair (1960), which the author contested then as "the Citroën *DS*" of Italian design, though still appreciating (then as now) its singularly high-quality design; by what – in the rigorous contraposition and the sensitive profile of the four open shells that composed it – bore witness to the 'architectural' derivation of the structures 'in view', though with an expression that Bruno Zevi would then define as more organic than Rationalist.

The problem (but not for the Castiglioni brothers) certainly did not lie in the quality of design, but rather the orientation that it suggested: that of the "de-syntaxisation" of the furnishing system. In this regard, the English critic Reyner Banham – besides his bizarre interpretations of design as pop art, reiterated in those years even during the ICSID congresses – had no hesitation in accusing Italian design of "having retreated from the modern movement"[14]. The future would prove to be even too explicit, as it might be said that the competitive procedure itself of the company system was what encouraged the abandonment of syntactic concerns by designers. In actual fact, the focus of design on the indi-

San Luca armchair by
Achille and Pier Giacomo Castiglioni
(1961, prod. Gavina, above)
and *Sgàrsul* rocking chair by
Gae Aulenti (1962, prod. Poltronova).

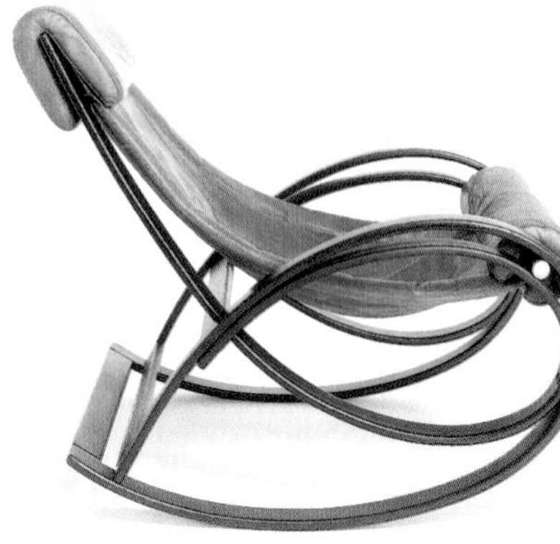

vidual object made the production of objects formally coherent with one another superfluous (from this viewpoint neo-liberty can hardly be defined as a style), initiating the final user to that 'taste' for formal contamination that would widely characterise furnishings in the last quarter of the cycle (1980-94) – making product sector specialisation possible and indeed ultimately embraced by enterprises.
At the same time (1962) as the appearance of the robust articulated structure of the *Sgàrsul* armchair (also defined as neo-liberty) by Gae Aulenti, the Castiglioni brothers began a period of apparently erratic and variegated semantic research that – by means of the Arco lamp that closely recalled the Ponti style – would culminate in the neo-Nizzoli volume of the *Spinamatic* beer dispenser (1964), which some would label neo-liberty as well, clearly ignoring the progress made: the semiologic step in the dignified transformation of a three-dimensional object into a communicative, commercial one – the truncation of the solid curve which, besides inserting the object into an ideal parallelepiped, highlighted the marked disk – and the structural one, of an auroral translation of geometry into topology. But despite the misunderstandings, the so-called neo-liberty style had the merit of shaking up the formal horizon even before its subsequent theoretical development, contributing to a process of aesthetical substitution by semiology that would later characterise the second half of the cycle (i.e. as of 1968), in its finest manifestations that prelude a new semiological sensitivity in products of the modern age. Umberto Eco[15] was to note that the dominant theme "of the second half of our century [...] has elaborated a totalising semiotic 'look', subsuming under semiotics even the problems of physics, psychology, biology and history", paradoxically adding that "much of modern semiotics seems to have set itself the task of sanctioning the end of its own object". A declaration that – decades later – would seem to attribute to semiotics its own life curve. Recalling the famous definition by St. Augustine – a sign is something that stands for another – then having examined the many sign categories, including iconic signs, *diagrams* (abstract), *drawings* depicting (concrete) objects, emblems and even targets, Eco once again resumed the thesis (1938) propounded by Charles Morris, who claimed that "something is a sign only because it is interpreted as a sign of something by some interpreter".
While on the one hand this view amplifies the meaning of the sign, on the other it specifies it only by relativising it to a particular pair of persons concerned: the *emissary* and the *recipient*. Such an interpretation, taken to its full consequences by Eco himself, will however attribute a new *status* to a work of art (and ultimately, to a certain extent, to any human work): that of provoker of an *interpretive collaboration* between author and reader, the very same status that Eco demanded in his narrative masterpiece *The Name of the Rose* (which by the way explains his explicit disassociation from the translation of the novel to movie). It is more than obvious that this way of considering (and therefore of devising) narration[16] concerns the designer of any object, of whom a new attention towards the semiological complexity of the object created is required. Of all Morris's semiological categories, the sensitive tool of information that went under the name of the "Que sais-je?" collection (Presses Universitaires de France) significantly witnessed in 1955 the publication of *la Semantique* by P. Guiraud (clearly inspired by the French-speaking Swiss Ferdinand De Saussure, 1857-1913), in 1975 *Signes, Symboles et Mythes* by L. Bénoist, and finally in 1985 la *Pragmatique* by F. Armengaud. It might be recalled that what might be called the

'lay' translation of the design concept that requires more or less free completion by the 'reader' would ultimately lead to that *customer satisfaction* which today seems to represent one of the front lines of corporate strategy[17].

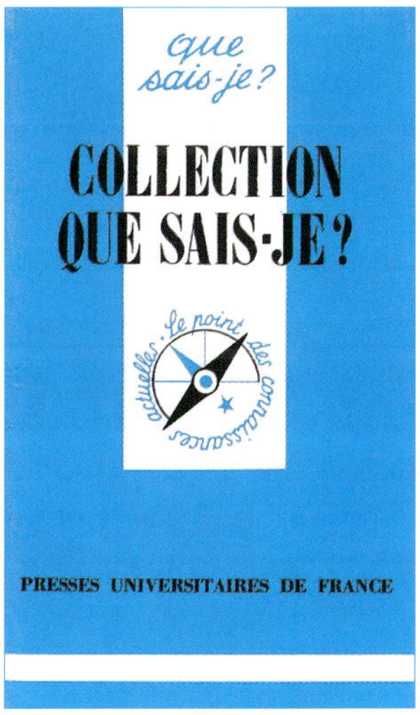

Catalogue from the "Que sais-je?" collection by Presses Universitaires de France.

The protest

In 1963, the constitution of the *Gruppo 63* – made up of philosophers and literary figures, mainly structuralists and phenomenalists, including Balestrini, Barilli, Eco, Giuliani, Guglielmi, Luperini and Sanguineti – seemed to announce a season of uncertainty, teetering between the avant-garde and the counter avant-garde, that according to some (including Eco) culminated in 1968 in a dead-end at the magazine "Quindici", of far different political significance. While Sanguineti maintained "a natural tension towards the avant-garde", Guglielmi claimed that "history is a lost value, no longer a meaning, just an occurrence" and Luperini had no hesitation in stating that "intellectual work is now completely inside the productive process" and that "culture is completely inside this society", a culture that "stands before us like an unshared object". On a vastly different front that might be defined as irrationalism, Elémire Zolla (1966) considered the use of the word "progress [...] repugnant" since science "does not tolerate humanism, extinguishes beauty, loathes contemplation" (sic), protesting against the consequent "uprooting of the vestiges of all aesthetic capability". And he went even further – unmindful of the lesson of Brecht's *Galileo* – to complain that "the superiors [have] thrown concepts that an erroneous interpretation can render dangerous to morally unprepared beings". Emanuele Severino would later (1972) oppose the condemnation of technology as manipulation and of science as a non-truth to the philosophy as the place of truth and salvation. Things that Franco Fortini – certainly not suspect – had branded as "ideological knick knackery", but which had, then and later, an important impact on the evolution, more than on the ideas (that the best soon abandoned), of attitudes – in this particular field, in the particular *Kunstwollen* that was to dominate the third stage of the cycle (1968-80). In his scathing pamphlet[18], the philosopher Paolo Rossi summarised the moment as that of radical opposition, *inside and out*, recalling the cautious sensitivity of Italo Calvino (1923-1985) who attacked the flight from rationalism – its *setback*, this Kierkegaardian term so frequently used in those years, partly due to the influence of Paci – as something typical "of venerable beards", warning that "there is at least one setback to reason every ten minutes; but the best thing is to see which bridge you are able to build to reach the other side and continue along your path. Only by this approach can new things still be seen as new [...]". This seems to the author the highest, most specific definition of the design project.
But the others were the premises – at least the Italian ones, so often associated with the ideas (suspect, and not only to the author, for political reasons) of Martin Heidegger – of a period not so much of irrationalism as of the irrational will for freedom from the bonds that were however there, before everyone, as heavy and unmovable as millstones. The end of the avant-garde and the avant-garde of the imminent end seemed to overlap in a Pirandellian pursuit of contradictions, usually expressed in non-falsifiable forms, which did not clarify where (and who) were those who had the good fortune to be called *apocalyptic* and *integrated*. All fashionable parlours hosted the game of *in* and *out*, to the point that it became *in* to be *out*, and *out* to be *in*. Such was the intellectual climate

in which design faced the second half of the cycle only recently concluded, and that would have its epiphany in the final quarter – totally unexpected by those who henceforth so proudly opposed *modernity* by their acts.

Italy, Bauhaus and Ulm

In the meantime in Ulm, Donau (Baden-Württemberg, FRG), 1952 saw the opening of the Hochschule für Gestaltung (HfG) which, at least in the intent of its first director Max Bill (1908-1994), would constitute the new Bauhaus, of which he himself had been a student. The HfG was also much discussed[19], though a summary would indeed seem justified, because of its importance (and consequences, also felt in Italy) as well as to provide a useful opportunity for reflection and contextualisation. First of all, its so-called 'internal' history began in 1943, with the execution – for high treason – of Hans and Sophia Scholl. Inge Scholl backed the foundation of an Ulmer Volkshochschule in 1946, in memory of his young siblings.

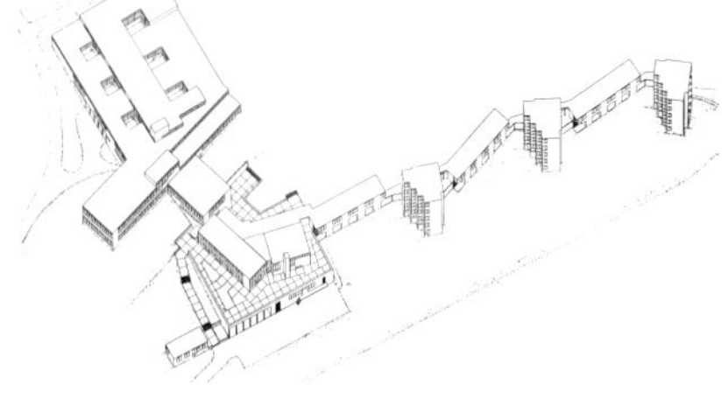

The Ulm Hochschule für Gestaltung building complex designed by Max Bill (1951).

Following the preparation of a program by Otl Aicher, Walter Zeischegg and Max Bill, the HfG – despite the immediate political difficulties – began in August 1953 its pedagogical activities in all the main fields of design, even beyond those covered during its time by the Bauhaus. The formal inauguration of the HfG in the new centre took place on October 2nd 1955 (with opening speech by Gropius), the year after Bill had become the HfG's first director (the following year the directorship was shared among Aicher, Bill, Gugelot, Maldonado and Vordemberge-Gildewart). In 1957 Aicher and Gugelot were awarded the Gran Premio by the Milan Triennale for the designs made on behalf of Braun AG of Frankfurt. The same year, Bill left the HfG due to differences regarding the orientation and syllabus; in 1958 the directorship was thus concentrated in the hands of Aicher, Gugelot and Maldonado. In 1959 Gugelot, though remaining in the school, abandoned it. Maldonado followed suit in 1961, though returned soon after, in 1962. In 1963 the school had once again a single director (Aicher), position occupied by Maldonado in 1964.

Leaving in 1966, Aicher was succeeded by Herbert Ohl; after 13 years of service, Maldonado finally left the school in June 1967.

In November 1968 the Regional parliament took the decision to shut down the HfG, the official reason being its financial situation. Less than 15 years since the Bauhaus, likewise for the HfG. Was there really any continuity between Dessau and Ulm? What actually distinguished Ulm from Dessau? In the catalogue quoted in the note, the Italian Renzo Zorzi – first director of the Edizioni Comunità, then key figure in Olivetti design management – pointed out the "risk of amalgamating personalities and positions" following the departure of the "survivors of the last Bauhaus" that pursued a "great ambition in a country whose intellectual tradition had not excelled in intellectual exchange [nor] in internationalism without preliminary issues". Of the stages of Ulm, well spelled out by Lindinger[20] (and attested through the author's own attendance) and its readings – to whatever degree they filtered through to the visitor (still close to Maldonado) – one cannot but admire the intellectual climate enjoyed inside the "convent" (Lindinger uses the same term). If continuity with Dessau was perceived in the latter season of Ulm, it was of a dialectic nature: as early as 1957, Bill was light years away due to the abandonment of the dominant themes of the twenties, teetering between Art and arts, technique and tech-

Graphic work by Andries van Onck, made at the Ulm Hochschule für Gestaltung in the course held by Tomás Maldonado.

nology, men and machines; in other words, because of the effective substitution of empirical technicality with the sciences, and consequently a new predominance of technology over technique. Despite the corrections along the way, Ulm was first of all a school of method. In 1968, Guy Bonsiepe, student and teacher at Ulm, ironically preceded his "*Arabeschi di razionalità*" in the magazine "Marcatrè" no. 41-42[21] with an aphorism by Flaubert (*Méthode: Ne sert à rien*) – and of this his protagonist must be proud, even though, following the demise of Aicher, Gugelot, Moles and others, we are the ones who cherish the memory together with all the teachers and students whose diaspora in Germany and throughout the world has proved so beneficial to design, even to the present day. With the warning that anyone sincerely wishing to know (or even practice) design of this century must inevitably pass through the tunnel taken by Ulm to overcome the barrier between the minor arts and industrial design. The atmosphere at Ulm was more that of a school of thought than of a school made up of teachers and students (the latter in fact as much protagonists as the former); the application of the ideas developed within the field of sciences – even those thought by too many as distant from 'concrete' design – was considered the way to identify the reasons to counter and even persuade coarse external 'realism' to take on truly pragmatic attitudes. But by the time dialogue could be established, it was too late; and not due to causes connected to a potentially different type of diplomacy, but rather to the general status of things, and particularly of politics. As much was clearly illustrated by the famous manifesto *Noi e il mondo delle merci*, which might be considered a stance on principle as much as a sign of the extreme and painful regret for the missed opportunities for collaboration. Ulm's fate was no exception to that of other great innovations, which demanded the existence, or at least the parallel construction, of a *system* inside which they could become tangible. Proof of this was Ulm's persistent return to the idea – not only, but also symbolic – of *environmental design*. Those who – from the right or the left (in this case difficult to distinguish) – contested what was undeservedly called the *Ulm style* even reproached the lack of attention paid to semiotics by the very school that had first consciously introduced the issue into design. Yet others – in Italy where Ulm had quite an important impact – go as far as to define its results as 'surrealist' (*sic*). The problem is that at Ulm, the semiotics were made up of *semantics, syntax* and *pragmatics*; for the new critics however, syntax was losing value, the relation between signs and meanings was melting away in a pursuit of senseless signs with the pretext of self-referentiality, while everything became pragmatic without the support of codes, except those used in the – parlour and not selfless – game of *in* and *out*. Certain aspects of the Ulm experience could prove useful when reflecting on the organisation of enterprises, at least in the relations that are established between *intent* (research), *will* (development) and the *implementation* of the ancient Chinese adage. Whatever the case – as a rereading of the entire series of "Stile Industria" by Rosselli bears witness – Ulm had a massive influence on the development of Italian design during the initial post-war period. Indeed at its closing (not only due to the definitive arrival of Maldonado in Italy, to commence, as irony would have it, his brief stint at la Rinascente), the legacy of Ulm had been transferred to Italy. And this also applied to furniture, despite the parallel influence of Scandinavian design that has, in some cases, preceded it.

Cover of issue 12 of the magazine "Stile Industria" (June 1957) edited by Alberto Rosselli.

Design and the technological "défi"

On the technical-economic front – still in the second quarter of the cycle – the thesis of the *défi americain* by Frenchman Jean-Jacques Servan-Schreiber, described in a fortunate, widely-translated book, resounded throughout the European continent. In Italy, bustling, lively discussions (sometimes even furious, such as seen at the Club Turati in Milan) were held on the widening *technological gap* that was moving the rest of the world away from the United States, threatening the possibility of what was then defined (probably for the first time) as *two-speed development*. In its basic truth, the thesis – corroborated by reliable data – was limited to the examination of the economic consequences of the *gap*, without actually specifying any immediately-feasible solutions. This became even clearer when Gino Martinoli – former Olivetti and Necchi executive, as well as cofounder/president of CENSIS, who recently passed away at the age of 95 – diagnosed the real reason for the gap as the insufficient number of technical personnel and managers that Italy was able to train every year, not to mention the methodological and qualitative shortfalls that prevented the Italian school from overcoming the barriers between the two cultures indicated by Snow (sciences and the humanities). Martinoli suggested the need for corporate intervention; not only through in-house training, but also by developing a thrust towards improvement and a multiple-discipline approach to basic educational training. His forecasts, including some of the quantitative ones, have unfortunately proved accurate for most of the remaining part of the cycle – indeed they were exacerbated by the conditions of the educational system (not the only cause, but particularly serious in Italy due to the 'deep slumber' of Italian institutional authorities) during the *protest period*, which to a certain extent continues. The social and business costs of insufficient basic training of managers and staff have not even be calculated, though they would constitute an effective warning. It was only late in the second half of the cycle that Italy began to witness the opening of authentic corporate training institutes (such as ISVOR under Marentino for Fiat and IAFE under Castelgandolfo for ENI), the modernisation of certain university training bodies (Bocconi university, the inclusion of microeconomics courses in certain institutes of technology), the creation of free universities (such as the LUISS of Rome by the Italian Manufacturers' Association). As regards design training, until recently kept separate from managerial training (given that qualified managers were not provided with information to contemplate its meaning, while designers received no managerial training), the ISIA was founded in Rome by Aldo Calò with the assistance of Argan (then followed by the institutes of Florence, Faenza and Urbino), while private institutes with 3-4 year courses were also being opened.

The social crisis – preceded by the student protests of the mid-sixties at Berkeley, even before those witnessed in France – punctually arrived in Italy halfway through the cycle (1968) in the form of the "*hot autumn*", then compounded by the *oil shocks* of 1973 and 1979. Deriving from (though delayed by) the same complex causes, these shocks would eventually lead to (among other important consequences) an acceleration in research and in the attempts to implement its results, obviously starting with those affecting the energy industry; the fact that oil-generated energy exceeded that produced by coal once again confirms the punctuality of the cycle. It was only after 1977 that energy efficiency – measured in thousands of 1987 US dollars of gross world product per ton of oil equivalent – would surpass the value of 2, then rising systematically to 2.34 in 1990. Interestingly, Japan and Germany registered levels higher than the USA both in absolute value and growth, mainly for structural reasons[22]. The thesis advanced by Servan-Schreiber, though in itself rightly provocative, proved in the long run (and *à la française*) to be more concerned with high technologies rather than those commonly used by enterprise. Reference to a similar fact (to be examined below) is made in recent EU data, which indicate a clear gap between the number of *scientific publications* per millions of US dollars spent – showing the USA and the EU followed by Japan, tailed by *New Industrial Countries* (NIC) – and the number of *patents* per millions of US dollars spent (which might be indicative of practicability) – in this case led by Japan and the NIC, tailed at a distance by USA and EU. A discrepancy that cannot but have an impact, at least in favouring product and process *development* in NICs, also on the basis of research conducted in countries that are unable to exploit their research. Furthermore, despite the development of "*technological parks*", *R&D* does not always lead directly to *R&D&D (Research & Development & Demonstration)*. This could and indeed should constitute one of the main sources of information for designers, when not actually an instrument for their training.

Another two macro-technological vectors were to burst to the fore, experimentally during the second half of the cycle, operatively during the last quarter: ecology (including, improperly but relevantly, the issue of the availability of physical resources) and electronics. The *question of the environment* – initially a braking influence, limited to production processes, while today concerning an entire scenario of technical and social implications – emerged as the determining factor behind a radical turnabout in the approach to product and process design within the prospects of *sustainable development*; electronics, on the other hand, set off an extreme acceleration towards *automatic process systems*, also allowing the integration of new control functions and the *miniaturisation of products*. While its applications in IT and communications assume a multitude of pervasive forms, to the point of being commonly indicated (inappropriately yet succinctly) by the term *new technologies* due to their predictable impact on all social, and even more importantly, military, organisations. The first political consequence of IT was in fact of a military nature, if there is any truth that, even before the political upheavals in the USSR, certain western politologists and polemologists claimed that a Soviet-like army would be unable to withstand others that were not only fully equipped with computer technology[23], but also had a civilian hinterland (civilian in the sense of consumer goods and even electronic games) with a population at minimum conversant with basic practices, when armed actually only requiring limited *specialised training*[24]. Bear in mind, however, that microelectronics were preceded by the last cases of mechanical miniaturisation, the most important and duly famous example being the *Grillo* telephone (Siemens-Auso) designed by Marco Zanuso and Richard Sapper, right at the end of the first half of the cycle (1967).

With a design horizon that was limited though offset by its great vivacity, furniture companies had begun to react to traditional symmetrical (and dull) formalisation of furnishings that was to introduce new objects – 'individual', 'niche' pieces of furniture that the Triennale had encouraged since 1954 – without actually producing archetypes, since they often remained isolated proposals, albeit alternative to 'formal' designs, such as in 1965 the CUB 8 (for Cassina) and the

Plastic bucket designed by Gino Colombini for Kartell in 1954. Below, two drawings for the patented process "for blowing bottles from thermoplastic material" (1954).

equipped, self-supporting wall by which Angelo Mangiarotti (for Poltronova) brought to new maturity the proposal made by Gio Ponti in 1949; or others by new companies, such as Bernini, Busnelli and Zanotta whose catalogues long boasted some of the leading designer names. In a lateral sense, Dino Gavina – another reputable designer – aimed directly at distribution, thereby leap-frogging production as early as 1954, with a frenetic, almost neurotic activity of cultural diffuser. If they all – in conjunction with enterprises led by important designer names like Gastone Rinaldi (RIMA) and Gino Sarfatti (Arteluce) – made a contribution to changing the furniture production landscape, striving towards a modernisation ridden with obstacles and diffidence (especially by a distribution network often ill-prepared to cope with these innovations), the companies supported by designer-architects had in Osvaldo Borsani of Tecno a paradigmatic yet peculiar example.

Besides the furnishings industry, the most important innovations witnessed from the beginning of the cycle concerned (with the exception of the first computers) functions and performance standards that could not have any significant impact on the general economics of physical goods – goods mainly designed to complete consolidated services, or innovations important in themselves and capable of modifying individual behaviour, but not enough to change the scenario of products to the same extent as the previous cycle. The *most significant genotypes* – such as automobiles, refrigerators, washing machines, radios and television sets – were, however, all in full development, giving rise to a concentration of design on their enhancement/refinement as well as on simplification/integration of functions, the replacement of structural materials (themselves undergoing complexification/simplification), on the review of processes and of what curiously has been defined by managerial consulting as *architecture of production* and the systematic reduction of final prices, for the main purpose of developing demand. It was only half-way through the cycle when a serious theory of *marketing segmentation* appeared that the practice of so-called niche policies would be used to a significant degree. These policies would have a preponderant role – on (and because of) a different way of approaching design – in the final quarter of the cycle (1980-94), in fact even to this day. If we disregard specific cases in large companies and in particular industries, designers were rarely involved as much as would have been useful in design projects attentive to cost considerations. What ensued was design (and the designer) construed as specific agents of high-segment strategies, when not of a niche market; an understanding that, while narrowing the gap between design and fashion, contributed to the confusion between *industrial designer* and *industrial artist* (according to the distinction made by the American designer/teacher Arthur Pulos, the former ICSID president, recently passed away).

In the meantime, design attentive to the values of the user-operator and his/her relations with the machine and the work environment was making its way into the same productive cycle through a broadened concept of ergonomics, adopted (one of the first in Italy) by Olivetti. After 1958, the protagonists of this theme included Rodolfo Bonetto (1929-1991), once lecturer at Ulm, as well as one of the first designers of his generation not to have come from architecture, but rather dividing his activities between *car design* and the *other design activities*. It might well be said that the existence itself of design not immediately intended for finished items had the merit of representing at least a factor of concern

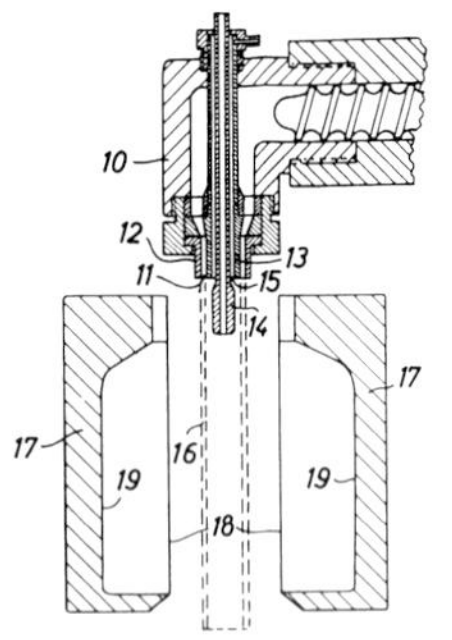

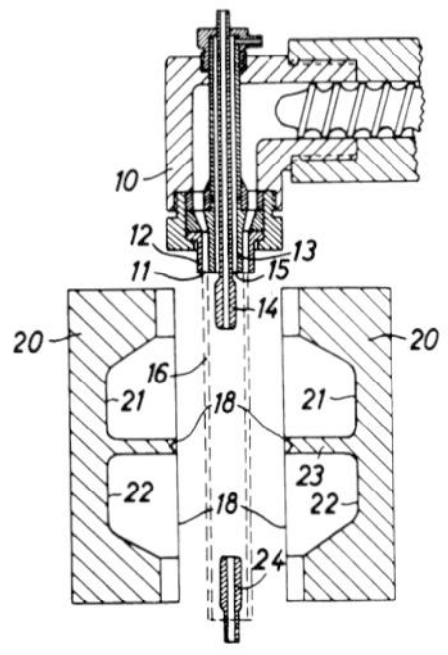

Above: armchair designed by Gastone Rinaldi for Rima in 1955.
Below: *Valentine* Olivetti portable typewriter (Ettore Sottsass Jr. 1969).

(thus food for thought) for those who had a purely aesthetizing and consumerist – in a word, hedonistic – view of design.
Before the end of the cycle in 1965, the US lawyer Ralph Nader published a book[25] that – denouncing certain serious shortcomings of industrial design – would influence policies governing industrial quality standards, especially at the beginning of the new cycle in which we find ourselves today.

Notes

1. Journalistically speaking, the dissent expressed in that era by various books penned by Vance Packard (immediately translated into Italian) should also be recalled.

2. We can give direct testimony after having known and worked with Luciana La Porta, in the early fifties secretary of the Olivetti press office (then directed by writer Libero Bigiaretti), who became world champion of speed typewriting, with over 450 strokes per minute, without mistakes. Luciana in fact had the letters on the keys erased because they were a distraction that disrupted the writing process. She claimed that the typewriter glided beneath her hands thanks to its solid curves, while also praising the excellent hold of the carriage lever. The speed afforded by the electric *Lexicon* easily exceeded 600 strokes per minute. As regards the formal peculiarities of the *Lexicon* and the *Lettera*, during a meeting of the Compasso d'Oro jury (late fifties), Ernesto N. Rogers declared that Nizzoli "knows things that none of us do" (author's own experience).

3. The list of Olivetti collaborators would be quite long. We mention in alphabetical order in Design Process, Olivetti 1908-83: Franco Albini; Gae Aulenti; Walter Ballmer; BBPR (Banfi, Belgiojoso, Peressutti, Rogers), Franco Bassi; Mario Bellini; Gian Antonio Bernasconi; Rodolfo Bonetto; Egidio Bonfante; Joseph Mueller Brockmann; Iginio Cappai and Pietro Mainardis; Michele De Lucchi; Egon Eiermann; Luigi Figini and Gino Pollini; Annibale Fiocchi; Jean Michel Folon; Franco and Simone Fortini; Jorge Fuentes; Roberto Gabetti and Aimaro Isola; Luigi Gabbrielli, Ignazio Gardella; Milton Gendel; Giovanni Giudici; Milton Glaser; Louis Isadore Kahn; Alan Irvine; Perry A. King and Santiago Miranda Camino; Le Corbusier; Albert Leclerc; Herbert Lindinger; Leo Lionni; Antonio Macchi Cassia; Vico Magistretti; Richard Meier; George Nelson; Marcello Nizzoli; Geno Pampaloni; Aurelio Peccei; Roberto Pieraccini; Giovanni Pintori; Carlo Ludovico Ragghianti; Arturo Rolfo; Bruno Scagliola; Carlo Scarpa; Leonardo Sinisgalli; Giorgio Soavi; Ettore Sottsass jr; George J. Sowden; Mario Tchou; van der Broek and Bakema; Pier Paride Vidari; Eduardo Vittoria; Hans von Klier; Marco Zanuso.
We add the following names up to the sixties: Giulio Carlo Argan; Giovanni Astengo; Antonio Barolini; Libero Bigiaretti; Anna Maria Bozzola Insolera; Francesco Brambilla; Luciano Codignola; Carlo Doglio; Armanda and Roberto Guiducci; Delfino Insolera; Anna Maria Levi Montalcini; Franco Momigliano; Franco Morganti; Luciana Nissim Momigliano; Ludovico Quaroni; Luigi Piccinato; Anna Maria Ortese; Franco and Michele Ranchetti; Ignazio Weiss; Bruno Zevi; Renzo Zorzi. Olivetti was in fact a veritable school for training a large number of management, consulting and technostructure executives in Italy.

4. Despite certain dissenting opinions (see Paolo Fossati, *Il disegno industriale in Italia*, 1970), Pinin Farina's work is to be considered a systematic example of structural and functional research (not only stylistic, not only of the automobile chassis) in a process of identification of a car shape that – beginning in particular with his partnership with Vincenzo Lancia in the thirties – led to an abundance of phenotypes (over 800 in 40 years of activity): consider for example the Cx of Cisitalia, measured in the wind tunnel almost ten years after the death of the great designer (1966). The result was a value of 0.37, unthinkable for the time (and only recently bettered).

5. As regards Kartell, see A. Morello and A. Castelli Ferrieri, *Plastiche e design*, Milan, 1984.

6. The first automobile made on a production line with a completely polymer chassis was the American Chevrolet *Corvette* in 1953. Only in the late seventies did IDEA (Turin) bring out a new model with a chassis made completely from plastic.

7. The exception of the most important text of the era – of international resonance – should be recalled: *Oligopolio e progresso tecnico* by Paolo Sylos Labini, first printed in a private edition in July 1956, then published in 1957.

8. Some of the trade names to appear on the Italian market over the next 10 years include: Ecat, Electrolux, Ermec, Fakir, Fargas, GEC, General, General Electric, Gibson, GRR, Homelight, Hoover, Hotpoint, Kelvinator, Klaver, Kobler, Krefft, Ignis, Invictus, Ligmar, Linde (AEG), Miele, Piccolobar, Progress, Promi, Rem, Riber (see Tersilla Faravelli Giacobone, Paola Guidi, Anty Pansera, *Dalla casa elettrica alla casa elettronica*, Milan 1989).

9. Significantly, the president of the Triennale of that era was Ivan Matteo Lombardo, social democrat, also exponent of the National Productivity Centre (Comitato Nazionale per la Produttività) that would later foster the Associazione Italiana per gli Studi di Marketing, founded in 1956 by Guglielmo Tagliacarne. The first general secretary of the Association was Giuseppe Ratti, later to become ENI executive, then ultimately president of the ICE.

10. The expo committee included the Castiglioni brothers, R. Menghi, M. Nizzoli, A. Rosselli, with A. Morello and M. Provinciali as organisers. The author was general secretary for the Conference.

11. The names include (without awards – over 230 in the 18 events held between 1954 and 1997): Franco Albini, Bruno Alfieri, Giulio Carlo Argan, the three Castiglionis, Max Huber, Gio Ponti, Ernesto N. Rogers, Alberto Rosselli, Albe Steiner, Augusto Morello, Bruno Munari, Marco Zanuso.

12. See "Stile Industria", no. 7, September 1996.

13. The first president of the ADI was A. Rosselli, followed by (amongst others) F. Albini, A. Bassetti, R. Bonetto,

A. Castelli-Ferrieri, G. Castelli, A. Cortesi, G. Iliprandi, E. Mari, A. Morello, M. Zanuso.

14. Indeed it is precisely Banham's thesis of design as popular art that cannot be considered extraneous to the subsequent – and far more radical – revolt against the Modern Movement.

15. See the item "Segno" in *Enciclopedia Einaudi*, vol 12, 1981. In 1975, Eco published the *Trattato di semiotica generale* (Milan), still a fundamental text of reference. More recently (and once again of great importance), *Kant e l'ornitorinco* ("*Kant and the platypus*"), Milan, 1997.

16. Umberto Eco, *Lector in fabula. La cooperazione interpretativa nei testi narrativi*, Milan, 1979.

17. As of 1960, the (often monographic) magazine "Communications" was published (Ecole pratique des hautes études – Centre d'études des communications de masse, Paris); its first editors included Roland Barthes, Claude Bremond, Georges Friedmann, Edgar and Violette Morin. The magazine's contribution to the study of scientific semiology – also in consideration of the attention to dozens of authors of international stature (and not only French writers, too often the case) – is incalculable.

18. P. Rossi, *Paragone degli ingegni moderni e post-moderni*, Bologna, 1989. The quote from Italo Calvino is taken from *Menabò*, 1963.

19. The basic text is *La scuola di Ulm, una nuova cultura del progetto*, 1987, catalogue of the exhibition of the same name curated by Herbert Lindinger, with the support of the Comune di Genova and Olivetti.

20. Herbert Lindinger, former student of the HfG, also underscored the internationalism of the school (as well as its antifascism) by recalling the quality and various origin of the lecturers. He subdivided its short history into six brief stages: the foundation, based on a new "pedagogical ideal"; the "new Bauhaus" stage, already characterised by the artisan-type, beloved artisticity that had constituted the unresolved theme of the Bauhaus; the period 1956-58, during which Maldonado and Gugelot "endeavoured to establish a new and substantially closer relationship among design, science and technology", intending "to tackle things more complex than seats and lamps", and with Maldonado, a "new form of fundamental course [...] increasingly [distant] from the Bauhaus concepts, [integrating therein] theory of perception and semiotics"; the period 1958-62, marked by the "euphoria of planning" and the prevalence of scientists and methodologists with the tendency to abandon the "universal concepts" to adopt a new "scientific modesty" with a "positivistic orientation" that favours the "hypotheses of work" over "programmatic manifestos", with methodology as the "doctrine of salvation", to the extent of bringing internal relations between teaching staff and students to crisis point; the period 1962-66, when Aicher and Maldonado made "an attempt to achieve a new balance between theory and practice, between science and design", ultimately defining the "Ulm model", though in the presence of an "uncontainable financial erosion", leading to the "final demands" of the Regional parliament; the last stage (1966-68), which "takes on the characteristics of agony" due to the attempts to change everything, including the political balance, and to lower objectives and dimensions. In the meantime the climate of the school had been heavily influenced by the radical demands of students in Berlin and Paris.

21. The editorial committee of "Marcatrè" for those issues included the art historian Eugenio Battisti (author of the *Anti-Rinascimento*), the musician Sylvano Bussotti, Maurizio Calvesi (former pupil of Argan), Gillo Dorfles, Umberto Eco, Roberto Leydi (later renowned for his studies conducted on student protest songs), architects Vittorio Gregotti and Paolo Portoghesi, the poet Edoardo Sanguineti.

22. Worldwatch Institute data, 1992.

23. Though to a lesser extent, the Arab-Israeli war was indicative of the technological and organisational inferiority of a technologically backward army.

24. Two centuries later, an apparent repetition of the causes for which the war technique of Frederick the Great, involving long training periods for each soldier, allowed him to win the battles but not the wars.

25. "*Unsafe at any speed*", Milan, 1967.

The Guzzini family: project and destiny

Cocktail set designed by Charles Joosten in 1959.

Development and growth in the Marche Region

If 1958 was a watershed year for the Guzzini enterprise, it was no less important for Italy, both for the significance of the political elections held that year and the maturing conditions offering new prospects for development, in the wake of the thirteen years of sustained growth that had characterised the initial period of reconstruction, through to the creation in 1956 of the Ministry for State Shareholdings. The period dominated by the growth of domestic demand was drawing to a close, while another was commencing, in which Italy – with growth rates that were even higher – was shifting investments towards manufacturing activities with new export potential offered by the fledgling Common Market. The divide between north and south in the country was, however, broadening, as shown by the abandonment of the land and the exodus that directed labour demography towards the industrial triangle.

These upheavals had a much smaller impact on the centre of Italy, including the Marche, despite the worrying loss of 100,000 of its inhabitants. As demographic data show, during the initial phase the most capable enterprises of the Marche had merely buffered the exodus, without actually stopping it by offering more employment. In little more than ten years however, from the early seventies the dynamic local business community had shown that it was more than capable of stemming the tide – even though it was arduous to distinguish whether the objective of employment was due to the structural conditions of the general economy or to the new opportunities offered by the market, given that the two often go hand-in-hand. The awareness of that objective – where present and *above* all if viewed with hindsight – seems to have been paradoxically distorted, in the light of the consequences (which alas persist) of the *capital* and *non-labour intensive* decisions made by the State Shareholdings (though here, once again, outside the Region). Nonetheless, as suggested by the testimony of the protagonists of the seventies and eighties such as the Guzzinis, it was clear that the latter concern lay at the basis of the main objective. Giuseppe De Rita could uncover in this twin purpose further evidence of his contention centred on *social cohesion* peculiar to Italy (that at least initially is not manifested through centralised facilities, regardless of the heavy investments made).

In the Marche Region, development and growth were in equal measure due to consolidated industrial enterprises and the expansion of semi-industrial companies (despite the fierce technological competition faced by some, such as in the musical instrument market, overwhelmed by electronic innovations, in particular by Japanese firms, countered tardily, as in the case of Castelfidardo). New companies also made their contribution, although not actually reaching dimensions comparable to established enterprises. Statistically, the Marche witnessed *development defeat growth*, though the latter was often the driving force (particularly as regards quality standards) that allowed the success of the former.

As is widely known (even here and today), small enterprises produce an overall degree of wealth greater than that generated by medium-sized ones, which in turn make more than the (few) large companies. While the limitations dictated by the topic of this volume preclude a detailed analysis, the Regional phenomenon of growth/development over those three decades – increasingly interconnected with the national and international framework – can at least be studied in broad segments.

Fungo table lamp designed by Luigi Massoni in 1965.

Regional industrial districts

The first segment regards industrial sectors and their role in the chain of production. As in the case of systems adjacent to the areas of agricultural production (though usually on a very small scale, mainly designed to cope with local demand only), the great majority of enterprises were oriented towards satisfying the requirements of nearby communities, though some were definitively projected towards more distant and complex markets. As a consequence, the goods types (and niche products) manufactured were locally interconnected, by the use of common technologies, materials or distribution channels, with constant exchange and opportunities for mutual learning experiences: this is the well-known phenomenon of the industrial districts already mentioned earlier.
Such was also the case of the Guzzinis, whose growth and development saw the prevalence of one main *rheology*: that of *plastomers* in sheet form, the production processes of which were explored in all directions (provided they showed promise for the properties and physical dimensions of components, in household, bathroom and lighting products). A similar approach was followed in the assignment of tasks to specific members of the family, in line with their natural flair – while also creating the catalyst for the internal and external enterprises, vertical in-house ventures or suppliers.
What was remarkable was that the personal connotations of each of the Guzzinis, though clearly evolving by dint of personal experience, proved to be relatively constant; in other words (based on hindsight), their qualities had been identified ever since the early years of the post-war period. In this sense, the family unit was never disrupted; it was rather a driving force, a unifying *deus ex machina*, present in the recesses of the conscience of each and every member, active beyond their individual assignments, behaviour and budgets.
This constant factor introduces us to the second segment – that of the strategies, the structures and the dimensions of the different enterprises that together were becoming a group, and consequently of the management and managerialisation of individual entrepreneurs (in the latter areas, the sense of Guzzini family solidarity led them along a path that ran opposite to the one taken by most other companies, which in fact tended to imbue a sense of enterprise in their managers). Firstly, by the immediate sense of order, that will become even clearer following the foundation of FIMAG, the family holding, though was not sufficiently influential to prevent cases of asymmetry and specificity induced by technical and market variations – and this also thanks to the specific personality of each family member. For example, a reading of budget or statement figures maintains all its strategic significance, not only nor so much due to the obvious differences in prospects that they reveal and justify, but rather to the mutual trust in the personal skills of the others.
This testimony is not intended to draw up an idyllic or hagiographic picture; anyone by profession aware of the tensions sometimes arising between managing staff (whether or not – or worse if – related) can distinguish the *functional in the pursuit of opportunities* from what is *authentically an organic component of the unit*. In this sense, the family does not appear as an entity to be revered, or a passive and preconceived factor of *solidarism*, as much as an active agent of *solidarity* – the difference between the two is not negligible. Giuseppe Guzzini, during the inaugural speech at the recent opening of the Historical Museum and Archive was thus able to unite the other side of the family who, having constituted the life force of Acrilux (acrylic component for light fittings), operate

ACRILUX

In 1963, Vinicio Guzzini and his brothers (children of Pierino Guzzini) incorporate Acrilux for the production (firstly by sheet, then by injection) of industrial and road lighting fixture components from thermoplastic materials.
Like other group companies, Acrilux paid great attention to innovation, in particular to production and process technology.

according to their own strategies along separate lines of development, while respecting the tradition of long-standing agreements of cooperation and non-competition.

1

2

3

4

5

6

"Per un'industria nuova" seminar, Macerata, December 1968.
1. Mr. Mazzini, vice-president of Farfisa, with Luigi Massoni.
2. Mr. Martarelli, director of Nazareno Gabrielli, with Rodolfo Bonetto.
3. Silvano Pierucci from P4 of Turin.
4. Mr. Ottoni, head of engineering department of Ariston Merloni.
5. Mr. Giuseppe Guzzini, administrator of Guzzini Fratelli.
6. Mr. E. Cecchi, sales manager of Ariston Merloni.
(from "Forme", n. 21, 1969)

Quality and project: design

The third segment covers the project and quality, quite possibly the basic strategy of the best-known and most important enterprise of the Marche, but in particular, and with great intensity, by the Guzzinis. To speak of *quality* (despite the fact that until the eighties the concept had a far more restricted, technical connotation than it was later to develop in the idea of "total quality") is to refer to an increasingly aware product policy – a discovery by which the Guzzinis progressed beyond the stage primarily oriented towards production and processes. Today the family still recalls the vital contribution made by Raimondo, Mariano's second child, who not only to the benefit of the company, but also to many others, as president of the Macerata business association, introduced in 1968 the issue of design during the first Marche convention specially for enterprises, attended by Rodolfo Bonetto (future president of the ICSID, the International Council of Societies of Industrial Design) and Luigi Massoni. That event merits special contemplation in order to understand its significance and purpose for the Guzzini family and the finest enterprises of the era alike.
First of all, design – as construed until then by academics, the most qualified designers as the most resourceful entrepreneurs – conferred cultural dignity to the rationalisation of the project/product as well as to the design process. The Guzzinis had already long given considerable attention to this aspect of pre-production corporate activities. The most fervent promotion of the entry of the designer into the company came from the versatility and culture fostered by Raimondo, convinced of the need to unite the "two cultures" advocated by Snow, though not to the socio-political consequences foreseen by the utopian Adriano Olivetti, nor within the abstract limits of formalism, such as the type practiced (with however great literary dignity) by Antonio Pellizzari in Arzignano (the businessman that Pavese had described in the *Diavolo sulle colline* was not only found on the hills of Piedmont). Unlike this type of restless entrepreneur, Raimondo belonged to a race that sat on the border between Catholicism and the active spirit of Weber Protestantism: the years of the transition from Roncalli to the modernist Montini can but support him. For Raimondo – as for a special type of other 'local' Italian businessmen – the two cultures were obviously in the family, though he knew that society (and the markets) now demanded a far greater degree of creativity, if necessary introduced from the exterior. An external creativity to be placed beside that of his older brother Giovanni, who felt that working in a company was like *penser avec les mains*; that of his younger brother Virgilio, with his cultivated technical flair; that of Giuseppe, who could translate values into numbers or this into that; that of Giannunzio, holder of patents for polymer transformation processes; or that of the youngest sibling Adolfo, renowned for both his innate and acquired managerial intuitions.
After a series of discontinuous partnerships, in 1962 Raimondo initiated a new design venture with Luigi Massoni. In this way design thus entered once again through the existing door of a humanising thought: Raimondo was not an engineer like his friend Giulio Castelli – who (he too heir of a family of coiners) had created Kartell in the immediate post-war period – looked upon by Raimondo

Trolley designed by Luigi Massoni (1969), awarded 2nd prize at the 1969 International Macef Fair.

with attention, exchanged with great esteem and friendship, as the testimony of Castelli himself bears witness. This despite the marked differences between the two: fideist and introspective the man from the Marche, lay and extroverted the Lombard. And yet both during the same years were flanked by a man: for Kartell, Gino Colombini through Anna and Giulio Castelli, for the Guzzini, Massoni through Raimondo.
They could hardly have been more different: two vocations and two temperaments, almost the complement of the qualities of their buyers. The Lombard Colombini, who arrived at design by instinct in the workshop of Franco Albini, rationalised everything he did with what one might call sceptical poetics, achieving unforgettable results; the Sardinian Massoni, near to 'purovisibilismo' of being *able to see*, played his skills on the sense of design, enthusiasm, instinct, attaining formal values that (later and after many others) his magazine "Forme" – whose existence and survival would also depend on the Guzzinis – would consistently display with few words but marvellous images that single-handedly brought out their very best.
Thus Massoni – coming from the Bottoni studio, attentive to Carlo De Carli of the Brianza renewal movement (no accident therefore that one of the most important guests was Alvar Aalto) – became a master of variation, as his works for Lagostina (saucepans) and Alessi (other domestic items) bear witness, as did the various Triennali in which he had a role that was far from secondary. His research for the Guzzinis was not limited to its own ends, but extended to the suggestions of other designers – from Noorda to Tilche, Belgiojoso to Bonetto and Lucini, to name only a few – thanks to the creation of a pool of external collaborators, articulate and constantly renewed (collaborating in addition with Giovanni and Raimondo in the construction of a product portfolio and the selection of distributors).
From the experience accumulated with furniture, Massoni – who has always straddled the role of designer, design-manager and (as they say today) 'communicator' – had also developed a certain sensitivity to the growth of Italian industrial districts. The development of the districts had coincided with the transition from individually-based small business to that of semi-mass production, and Massoni's experience in this regard certainly had not gone unnoticed when Raimondo and Giovanni drew him away from Andreani, for whom Massoni, still at university, edited the ICI "Notiziario Perspex". For the Guzzinis then, the years between '64 and '74 could be defined as the "heyday" that preceded a more specific commitment to the study of applications with new materials – such as ABS – that, despite the enormous technical skills shown by polymer producers in Italy, never actually constituted part of their competitive strategies. The author himself was witness to as much, having worked on the strategic themes of the years between 1970 and 1980, when an important game was played; a game that was lost due to the inability or the refusal to perceive research as a key factor of marketing. The Guzzinis – by the testimony for example of Matiussi of Montedison, the protagonists of research efforts that struggled to find strategic dignity – tirelessly urged the production of materials that *ante litteram* could be called "customerised". Their commitment went even further; through the new polymers, in the wake of the hackneyed history of methacrylate for aeroplanes, they were amongst the first to press on that *technology-switching* (e.g. from the automobile to household products, and vice versa) which would after 1975 constitute one of the dominating issues of development, in

The Picnic stackable plate by Rodolfo Bonetto, designed in 1967 and produced in 1971. Made in a single piece by injection moulding.

1st Macef prize awarded to the party service "Modulo Quadrato" designed by Ornella Noorda. Amongst the persons present: Raimondo Guzzini, Bob Noorda, Luigi Massoni, Ornella Noorda, Adolfo Guzzini, Fabio Lenci. Milan, September 1968.

Italy as elsewhere, when the problem of *energy-saving* was seriously raised. Secondly, and even before the years of street protests inspired by 1968, opinions that contested the idea of the simple functionality of formal qualities had surged to the fore. One of the merits of the Guzzinis was the elaboration of a corresponding strategy that was rationalised in later years. It was indeed in those times that design assumed the role of protagonist; but for the Guzzinis, a protagonist to be helped so that it could eventually help itself. In the words of today's generation at the head of the various Group companies – created precisely to give the products their own specific, economically-sustainable quality – it is clear how design has since become a strength to be applied to each product type, but also something distinctive to be handled with care, like a family hallmark.

Two witnesses

Giulio Castelli

Giulio Castelli.
Pupil of Giulio Natta, founder and today honorary president of Kartell.

"My acquaintance with the Guzzinis started with Raimondo, we were good friends; Kartell began producing household goods, then together with Guzzini we became the most important manufacturers of plastic household articles as of the mid-fifties. Right through the mid-sixties, when Kartell diversified its production, we were actually competitors, though in a different way; they mainly produced tableware, while I made mostly utensils, buckets and other kitchen accessories. Distribution was, however, the same.
I was close friends with Raimondo. We would meet at the fairs: at the Milan Macef or in Cologne, we'd exchange information. The Guzzinis were specialised in applications with Plexiglas and methacrylates in general, and in vacuum forming. We too made methacrylate articles in our *vacuum forming* department.
In terms of goods type, the competition between Guzzini and Kartell was broader, because it also covered lamps and furniture, which the Guzzinis manufactured for a certain period. We were also competitors in the lamp market, because Kartell had produced some models that it sold in the furnishings sector, then we continued in that direction. Lamps were our key into furnishings, we produced them with injection moulding technology, a technological process that the Guzzinis didn't dare use. I've never avoided any type of plastic material, and I used any type that suited my interests; but the Guzzinis were too 'methacrylate-oriented'. We were a kind of interface with materials. We competed fairly, also because we were in the same market though with quite diversified products, even in terms of production technology."

Gaetano Clementi

Gaetano Clementi.
Graduate in 1946, he collaborated with Fratelli Guzzini during the fifties as consultant for the production of acrylic plastics.

Says Gaetano Clementi, founder in 1959 and present-day manager of SimaPlas, a Guzzini privately-controlled company: "Pomezia [where the plant was located] was the site of the Dino de Laurentiis film studios. We came into contact with American technicians who used methacrylate in their production designs. "Barbarella" was for us a fundamental experiment: the creation of a large sphere with a two-metre diameter, containing more than two cubic metres of water (in which Jane Fonda swam naked) was a real enterprise, in fact we had to give a written guarantee against the occurrence of any problems. We also built the enormous transparent dance floor, illuminated from underneath, that was later used in all sorts of venues during those years. Once again, we had to provide guarantees that it would not collapse or break: the insurance cover ran into millions. For the floor, T-shaped interconnecting sheets were made, supported by transparent cylinders. The person in charge of production design was the architect Mario Garbuglia, and all work was closely followed by Virgilio Guzzini. In any case we asked to be present at all takes, to make sure that the structures were used correctly – but we were also curious to see Jane Fonda. We were also later involved in the production of sets used in "Doctor Faustus", starring Liz Taylor.

We worked in the production of special articles for movie sets for a few years, after 1967; we were present in almost all the movies with the sets designed by Garbuglia".

Unity in diversity

Referring back to the interest in new materials discussed above, it is possible to note a logic shared by the strategies of appearance and those of structure, of form and performance.

The recognition of an *ethos*, so often transversal, is the explanation behind the birth and development of a Group united in its articulate diversity, in the flexibility derived from the determination to identify and solve problems by a common method, of exchange and participation, sometimes opposed, always discussed.

The *synchronic* aspect of this ethos is characterised by the principle that anything concerning the product (the Guzzinis were in fact strongly, even exclusively, *product-oriented* until the late seventies) overrides all other variables, including marketing; technology and economic performance remained at the heart of all innovations. But this also determined the development of a special ability to combine technical bonds/opportunities with economic constraints, giving rise to product configurations characterised by a constant realism towards the user, something that certain people could mistake for mere formalism. On occasions the products became expressions of the production process itself, if not of the interest in the specific nature of the material used, in line with the vocation and ability of all to translate it into object.

Plexiglas spiral staircase, Polivar offices, Pomezia.

Which leads us to the second aspect of the Guzzinis: the *diachronic* one.

In the evolution of all acts of implementation, innovation is read as something that is not an isolated act due solely to the creativity of designers (and others), but rather a complex phenomenon that accounts at all culture stages – general and daily – for the system in which it appears. The Guzzinis too passed through the design stage by which, in the effort to co-opt the final user himself, only received any *feedback* through distribution; in the more advanced stage, a proclivity towards innovation would emerge, characterised by an even greater adherence to the current cultural moment in accounting for the real conditions of competition. To differ so that the product communicates a different and peculiar way to construe and use it, but to differ just enough to not require the guidance of classes considered superior because more cultured, clever and informed. For this reason, Guzzini products have rarely been proposed to the *elite* with all their neuroses (and smugness); they rather address the real, current consumer directly, trusting in the ability to take the innovative step necessary, as if to accompany him or her to choose Guzzini over all others.

And yet – astoundingly, especially in the light of testimonies given by leading designers, such as Angelo Cortesi when he was president of the ADI – no research, however difficult, costly or even unlikely has ever been excluded on principle. This openness would have, particularly as of the seventies, given results – in lighting, sanitary articles and even household products – that translated individual products into systems that gradually, almost naturally, became *services*.

Suspended *Disco* lamp designed by Cesare Casati, 1966.

Research and innovation

In actual fact the Guzzinis seemed to construe research as a forecast of what could (and not necessarily should) be developed (and must therefore be known), more than an immediate application; the market is to be confronted just before the time in which a demand sufficient to vindicate product, production and distribution costs is predicted. Subsequently, an attempt is made to anticipate it through investment in communications that determine it. Research invents, innovation applies.

Investment was therefore firstly concentrated on design and estimation costs without which there is a risk of not knowing where to proceed, but at the appropriate time; secondly (especially in the last decade), it centres on targeted communications next to final demand (*promotion*) when products are addressed to satisfy it, or recently by advertising, on communication appropriate to the least known aspects of the *contract* – through to, for example, the recent slogan used by Guzzini Illuminazione "Who stole the Milky Way?", which bears witness to the intelligence of the increasingly widespread concern over light pollution in public and collective areas.

The eventual result was a series of product portfolios (one of the broadest yet most target-oriented around) that – at least in the items directly designed for the end user – seem to vindicate Banham's contention mentioned earlier that saw design as a form of pop art (provided we know what the term actually means today). In the light of the Guzzini experience, the contention becomes even less singular – indeed one might even say that the Guzzinis, even more often than others, have succeeded in incorporating marketing into product policy without making design a mere factor of marketing.

We thus witness a case of original *marketing orientation* that, begun by instinct, has gradually become more aware and rationalised; an orientation made up of sensitivity shown initially in the choice of designer, then in gleaning (not without prior discussion) the final outcomes – a process involving extensive in-house participation to achieve corporate consensus, essential to overall success. The designers themselves often refer to the enormous respect, even reverence, that the Guzzinis have for creativity, if for no other reason than their own assiduous experience in the field, allowing them to assess and appreciate the creative effort of others. They are therefore more authorised, more facilitated in criticism, which has progressively shifted from a more technical approach towards one more closely related to sociological imagination that often anticipates field research, in the real market.

Medusa table lamp designed by Buttura-Massoni, 1965.

This slow, systematic turnaround in approach brought about, over time and as "an innovation in continuity", an evolution in choices, by sector designers themselves through to those regarding their proposals. After Massoni – a leading player in the Guzzini adventure – the list goes from Gio Ponti to Ambrogio Pozzi, from Paolo Tilche to Fabio Lenci, from Bruno Gecchelin to Furio Minuti, from Piano to Lovegrove and Norman Foster, each with a different task. And what these names and others have achieved beyond the projects set aside would seem to corroborate our old idea: that the history of design is the history itself of its definitions. Indeed it could not be otherwise for the Guzzinis, given the array of ambitions and products, some fulfilled, others rejected; the functions and roles demanded of designers during the various stages of company history. But apart from the minor, negligible components, one perceives that the array is by no means an accident, but rather bound by a red thread spun by the many,

Above: *Pic Boll* picnic set (Fratelli Guzzini Studio, 1978) in a sales brochure published in 1981. Plates and cups are stored inside two hemispherical containers that double as salad bowls.
Below: *Pasticcera Bonetto*, pastry tray designed by Rodolfo Bonetto, 1970.

successive stages of the endeavour to be adaptive and aggressive, ready to integrate the quality of the moment while giving it its own, singular, original interpretation.

Until the second half of the eighties, the Guzzini product strategy had a constant feature: a narrow path that runs between low-profit certainties with guarantees for the immediate, and the risky proposals for the short term, though promising over the medium. As a rule, Guzzini products have their own individual structural and formal features, because something in the new is borrowed from the average, beyond the personality of each designer; properties that reveal a sensitivity for large-scale diffusion: sections and thicknesses greater than the normal, to assure solidity even at the price of greater material requirements or additional bracketing, but sleek forms to simplify structure and accelerate the transformation process. The evolutionary can be read in the increasingly high number of curvature radii by which products are distinguished from both the comparable ones in the distribution assortments and the ones encountered in end use: a way of bringing *consumer and user* together, just when the distinction between the two was so widely discussed and maintained by critics in the late seventies.

Marked by a sequence of products rather than a mere *defilé* of projects (and designers), the history of Guzzini design is fully identifiable with the history of 'real' design. In other words, anyone writing the general history of design and wishing to avoid running after the various avant-garde movements, could look to the Guzzini catalogues as a repository of design that has been effectively welcomed – at least in relevant sectors – by real users. The weight of this long, ongoing adventure has had an incalculable impact in the consciousness of design (at least in Italy): in the author's opinion, just as important as other proposals, and not always particularly courageous at that. In fact, given the shapes and quality of products, it is not always possible to distinguish the prevalent feature of the result: whether it is the determination to assert the personality of the enterprise by means of a design perceived as entrepreneurial resources, or the aspiration to modernity through a design perceived as a means of cultural dissemination.

The choice of materials and diversification

Even before the crisis that hit in the second half of the sixties, the opening of Harvey by Mariano's sons had posed new problems of diversification (to which reference has already been made when discussing the introduction of new materials and creative propositions). At least in intent, plastic could no longer be seen as the sole protagonist; other solutions with a less Calvinistic choice of materials needed to be explored, also according to the specific performance expected of each object. It was in any case during the development of Harvey that – as further evidence of the parallel development of design, structure and process – collaboration commenced, for lamps, with Fabio Lenci, a young Roman architect (b. 1935) of Marchegian father. That short period of time saw Lenci involved in many different tasks and roles, showing great attention to new materials on the one hand (such as foam and hard polyurethane), while on the other to new light sources that had begun to appear between the sixties and seventies, such as iodine lamps, a technology that saw important developments at the hands of Lenci himself. Lenci's engagement with Guzzini was to open a new

The first round shower, designed by Fabio Lenci in 1972.

phase in 1970, the year when Virgilio convinced the family to become involved in the new initiative that, at the cost of great perseverance and commitment, was to become another source of company pride: Teuco.
Teuco was incorporated as a company for the production of health care equipment. Starting out with the specific intention of replacing all materials traditionally used in the bathroom with polymer-based material (such as the famous *round shower* cabinet), the facilities became increasingly advanced, combining functionality, *comfort* and enjoyment.
In this activity, Virgilio and Lenci (who continued his involvement in the Guzzini lighting sector; in 1973 Giovanna Talocci joined his studio) were able to express the finest of their talent, while keeping faith with the now-established approach of designing, developing and producing objects that the user would want to posses, but with a subtle form of discretely-introduced creativity, accompanied and backed by thorough intermediate- and end-market studies.
The two were also jointly active in the fields of advertising and catalogues, once again confirming another Guzzini tradition: that of not limiting designers to mere design, but involving them in the compilation of the portfolio and the communication of project significance to the exterior world.
Speaking about the subsequent introduction of the whirlpool tub, Lenci may have been oversimplifying when he quipped that not only its inspiration but also its specific structure was copied from a competitor; but even if this had initially been the case, it would have been an eminent example (because vindicated by success) of *creative destruction* – in other words, of what Alois Schumpeter specifically attributed to the entrepreneurial function.
The competitor in question indeed managed to draw away one of Lenci's collaborators, and another idea soon emerged: that of the home sauna, made with simple but decisive experiments, like the ones conducted by Ravenna businessman Serafino Ferruzzi, when he tested his shallow-draft transport boat models in his bathtub; once again, cases of authentic creativity, not far from the first experiments conducted by Volta or Galvani. The totally original idea for the introduction of ultrasounds was clearly different. The idea underlined a certain continuity of creative behaviour, i.e. of the maxim that one idea gives rise to another, and one that was immediately entrusted to skilled hands, in this case, Lenci for the design. Other Lenci projects were characterised by an approach to experimenting comparable to the pursuit of the avant-garde, as discussed earlier.
The collaborative effort by Bruno Gecchelin (b. 1939, Milan) with Guzzini Illuminazione dates back to 1976; his design for the *Shuttle* a decade later (1987) was awarded the prestigious Compasso d'Oro prize (15th year, 1989).
Gecchelin came from years of experience in leading furniture manufacturers in Lombardy and a stint as designer for Olivetti (with Sottsass), periods during which he had developed a profound interest in the integration of technology with design. He was the right man for Adolfo Guzzini, who indeed convinced Gecchelin to become involved in a line of design projects in which innovative performance went beyond the initial decorative facets to override all other considerations. In one way he overturned the conventional logic, insofar that he was favoured by a prevalently collective, more diverse and rational recipient, as well as an unprecedented openness of the private consumer to the proposition of *hi-tech*. The proposition was more important than adaptability; a line could thus be constructed on the basis of an autonomous, aggressive policy, in the face of which the appearance of other new lighting sources – such as the dif-

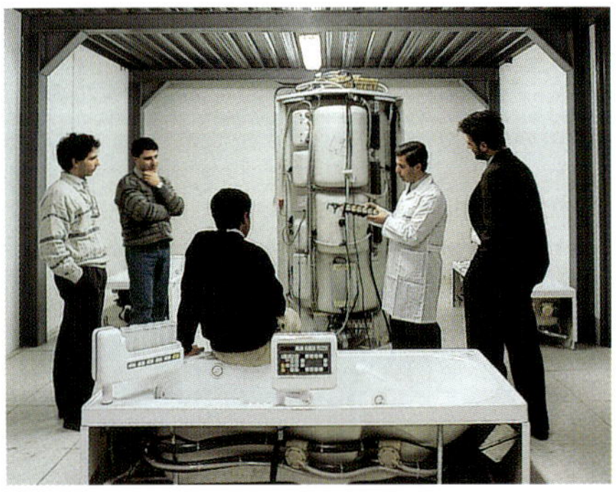

Production of electronic components for whirlpool tubs and shower cabinets at the Teuco factory.

ferent type of halogen lamps, or the new fluorescent types – merely accentuated the need to propose/impose than to blandish. 1979 was the year of the great leap forward, triggered by the research organised by Adolfo, who thus acquired external experience (as in the case of die-cast components), with the need to decide what was best done internally, particularly when high-risk investments were under consideration. The passage separating certainty and risk therefore tightened, to the extent that risk became a certainty, but the only one able to remunerate the work of all.
Following the remarkable acumen shown in selecting the right time to 'come forward' (and so often just a moment beforehand) shown in 40 years, the Guzzinis – starting with light fittings and Teuco, though already expanding in the other corporate components – took on a new, voluntary role: that of explorers of what the end user could really desire. What might be defined as the new family 'vocation', which has now been embraced for many years, was correlated to the timely intuition of the neo-Schumpeter revolution in business: the most rapid, radical change demanding of enterprise greater risks, courage and rationality, therefore far more flexibility and creativity on the basis of more extensive knowledge to be systematically acquired.
The *intuition of the future* was thus being doubled: one part rooted in the *ex ante* phase, when investing in research (by then allocated substantial investment), the other *ex post*, when checking that design complies with the technical, structural and performance criteria specified.

Two witnesses

Antonio Renzi

Antonio Renzi.
Graduate in Economic and Commercial Sciences from the Università la Sapienza in Rome; following a management training course at the Ancona ISTAO, he joined Teuco, of which he is today Managing Director.

"When I started working with Teuco in 1973, the sanitary ware market addressed by the company was a typically mature, traditional one, both in terms of the materials used and the designs – after all, the articles had a simple functional role to fill. The Teuco idea, quite revolutionary for the time, was that of furnishing the bathroom with greater emphasis on aesthetics, the role of colour and coordination.
The success of Teuco products – characterised by clearly differentiated features – right from the beginning showed that the traditional items present on the market and offered by the leading names of the period no longer corresponded to the new needs of consumers, who were becoming increasingly refined and discerning.
In 25 years of business, the company has followed the evolution in market demand, and in fact has made an important contribution to changing the type of products that people look for. The main changes are, in products, an increasing demand for matching, coordinated articles, with more functions, in response to the rising standards in the quality of life, relaxation and well-being; people also demand products that reflect a certain status. The increasing complexity of products has led to the acquisition and implementation of new technologies and skills, starting with electronics through to the recent application of ultrasounds in the whirlpool system.
As for the more immaterial demands, people look for a higher standard of service, pre - and after-sales assistance, more attentive relations between the distribution network and the end consumer.
In communications and advertising, a need has emerged to highlight the more emotional side when using whirlpools and showers, on total user gratification.
Teuco has witnessed 25 years of market transformation and evolution, following a specific strategy: that of differentiation in its fullest sense. Differentiation of product, of service, of organisation, of communications."

Cabriolet minisauna (1980) designed by Carlo Urbinati.

Antonio Santi

Antonio Santi.
Graduate in Economic and Commercial Sciences at the Università la Sapienza in Roma, following a management training course at the Ancona ISTAO, he moved to Recanati to enter iGuzzini illuminazione, of which he is today Managing Director.

"I became acquainted with the company in 1974, as a student at the ISTAO when I was writing my thesis, thanks also to the formidable openings and relations that existed with the university.
In 1975 the company closed the year with a big loss, but even then, at critical times, it looked to the future by insisting on the induction of young workers; and that same year I was called for to join them.
The crisis had been brought on by the oil shock and the spiralling prices for plastic materials, which seriously disrupted company plans.
A series of operations was therefore conducted to ensure that – despite the cuts to corporate turnover – on the one hand the books were balanced, thanks also to cuts in expenditure, and on the other the launch of technical lighting products.
The sheer enthusiasm in the thrust towards these objectives, as epitomised by Adolfo's total commitment, helped us to rapidly overcome the crisis and take the company to where it is today: European leader.
The company grew swiftly, becoming increasingly complex, so important organisational changes were also necessary. We immediately turned our attention to creating in the corporate organisational structure a level of intermediate managers who assumed their own responsibilities, so that it would feel part of an executive group. This also required that the entire company recognise the role of the group as a type of transmission belt between the top and the foundation of the company. The process benefited from the completely innovative style and approach to personnel training adopted.
Changes were first made to the technical departments, in particular design, then in the eighties it was the turn of the marketing office.
Of special interest was the lighting project office created to offer architects and distributors a consulting facility for the development of lighting projects in line with the latest design criteria and technologies available. In terms of human resources, this development increased the proportion of professionals involved in research, dedicated to innovation in marketing, distribution and technology.
The upshot of this strategy can be seen in the high training standards already boasted by company personnel as of the late eighties.
Today's objective is to bring the company even closer to the customer by exploiting all IT and management systems available. This is our main organisational project as we work towards the future.
In actual fact we already use our IT facilities to coordinate from the head office the main activities of all foreign branches and peripheral sites.
But our basic strategy remains fundamentally the same as always: on the one hand, product innovation to meet market demands; on the other, investments in human resources, with emphasis on team work and unity."

Interviews by Anty Pansera, taken from the Archivio Reale/Virtuale of Guzzini companies. Recanati, September 1999.

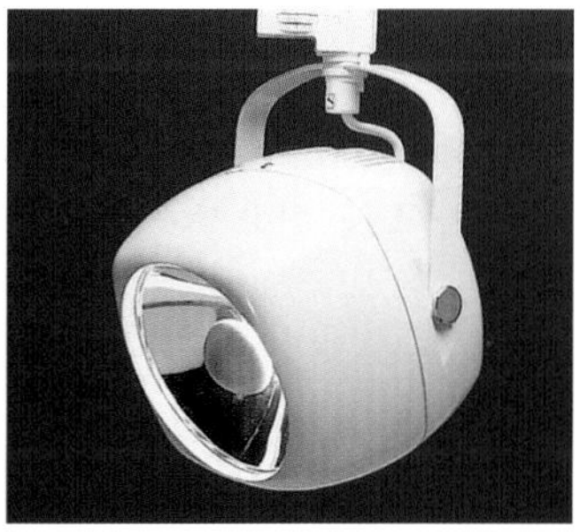

Ufo spotlight designed by the iGuzzini Studio in 1977.

Design and the evolution of a managerial culture

The management of the new process of research/design/production that was emerging required a corresponding heightening of managerial quality; since then, design has become an authentic challenge in the pursuit of a synthesis more of concepts than of form, and more of form than of formality (or worse, formalism). Clearly the transition involves an elevation to a higher level of abstraction, which can only be defined as the *design of design*.
We should not however be deceived by the term "abstraction", because it is precisely due to and by means of its higher level that the appropriateness, even the variability, of a design can be assessed, along with its apparent formal inconsistencies, its flexibility, and therefore its concreteness. The experience of a user segment can easily shift to satisfy the requirements of another, thus multiplying the revenue deriving from the original research. So design becomes research and research design, jointly contributing to the bonding of the two souls of business: the one that revels in tackling risks, the other more cautious, tempered

The North Greenwich Transport Interchange in London, designed by Norman Foster and illuminated by *Greenwich* system by Norman Foster, 1999.

by the certainty that expertise is best used when the effort is made to acquire it. During the acceleration of technological and sociological dynamics, this change in level became necessary, even urgent; but not everyone was able to make the change, and today pay that price. A high price, for the many who had not perceived the need for change (and who are consequently vanishing from the area of economic exchange), and the others who, positioning themselves between the pursuit of a supposed freshness of ingenuousness and its clever exploitation, have let themselves go to the pointless metaphors of the decorative stage (paradoxically risking an early exit from the area of cultural exchange).
For the Guzzinis, the new policy of "design of design" coincided with the definition and expansion of partnerships with external collaborators: the designing process is performed seamlessly from the very first stages, mainly comprising basic research and the development of its applications (by highly-specialised personnel), to the intermediate stages involving feasibility tests (mainly in-house), through to final design targeting. The process is in fact becoming increasingly complex due to constant product return and re-examination, as well as the pressing need to slash *time-to-market*. For all these reasons there is a clear orientation towards *concurrent engineering* wherever possible. New, important names and personalities are making their presence felt: Piano, Lovegrove, Foster.
The most recent Guzzini creations demonstrate that, in all their goods sectors (including those covering older products, such as those for the household), the elevation to a higher quality standard has been made, even though the same continuity of production remains. Indeed the fact that the concept of "design of design" has been embraced in no way contradicts the principle, still faithfully observed, that we learn by doing – and even doing not only what is intended, but also encountering the "blessed moment of good fortune", the unexpected, *serendipity*.
The Guzzinis have thus implemented a general strategy that seemed to only befit corporate groups much larger than theirs: that of the culture of *managing different cultures*, in both the specific object of interest as well as the advent of new generations. Towards the latter, the older generations insist on the need to possess a common method for managing different areas, able to perceive their specificity while capitalising on not only (and not so much) the overall dimension, but also on the growing complexity of internal and external expertise.
A reading of products shows (better and more than any general comment) how and to what extent the synchronic and diachronic aspects of the Guzzini ethos have dominated the scene of their own evolution, while having set a standard for others. Besides explaining why the Guzzini name has always represented the other side of Italian design.

Men and issues of innovation

Marco Migliari, Dario Moretti

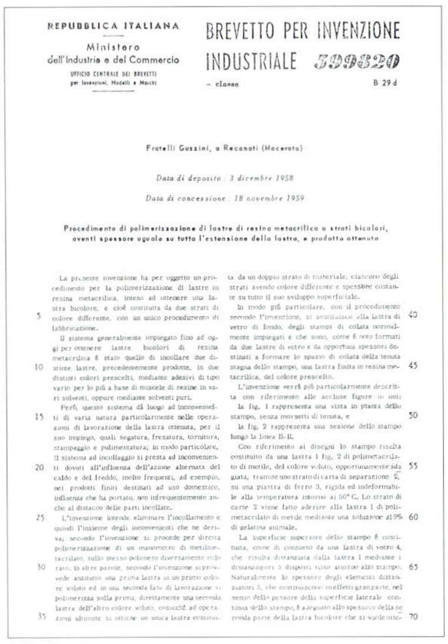

REPUBBLICA ITALIANA
Ministero
dell'Industria e del Commercio
UFFICIO CENTRALE DEI BREVETTI
per Invenzioni, Modelli e Marchi

BREVETTO PER INVENZIONE INDUSTRIALE 599820
– classe B 29 d

Fratelli Guzzini, a Recanati (Macerata)

Data di deposito: 3 dicembre 1958

Data di concessione: 18 novembre 1959

Procedimento di polimerizzazione di lastre di resina metacrilica a strati bicolori, aventi spessore uguale su tutta l'estensione della lastra, e prodotto ottenuto

La presente invenzione ha per oggetto un procedimento per la polimerizzazione di lastre in resina metacrilica, inteso ad ottenere una lastra bicolore, e cioè costituita da due strati di colore differente, con un unico procedimento di fabbricazione.

Il sistema generalmente impiegato fino ad oggi per ottenere lastre bicolori di resina metacrilica è stato quello di incollare due distinte lastre, precedentemente prodotte, in due distinti colori prescelti, mediante adesivi di tipo vario per lo più a base di miscele di resine in vari solventi, oppure mediante solventi puri.

Però, questo sistema dà luogo ad inconvenienti di varia natura particolarmente nelle operazioni di lavorazione della lastra ottenuta, per il suo impiego, quali segatura, fresatura, tornitura, stampaggio e pulimentatura; in modo particolare, il sistema ad incollaggio si presta ad inconvenienti dovuti all'influenza dell'azione alternata del caldo e del freddo, molto frequenti, ad esempio, nei prodotti finiti destinati ad uso domestico, influenza che ha portato, non infrequentemente anche al distacco delle parti incollate.

L'invenzione intende eliminare l'incollamento e quindi l'insieme degli inconvenienti che ne deriva; secondo l'invenzione si procede per diretta polimerizzazione di un manometro di metilmetacrilato, sullo stesso polimero diversamente colorato, in altre parole, secondo l'invenzione si provvede anzitutto una prima lastra in un primo colore voluto ed in una seconda fase di lavorazione si polimerizza sulla prima, direttamente una seconda lastra dell'altro colore voluto, cosicché ad operazioni ultimate si ottiene un unica lastra costituita da un doppio strato di materiale, ciascuno degli strati avendo colore differente e spessore costante su tutto il suo sviluppo superficiale.

In modo più particolare, con il procedimento secondo l'invenzione, si sostituisce alla lastra di vetro di fondo, degli stampi di colata normalmente impiegati e che sono, come è noto formati da due lastre di vetro e da opportuni spessori destinati a formare lo spazio di colata della tenuta stagna dello stampo, una lastra finita in resina metacrilica, del colore prescelto.

L'invenzione verrà più particolarmente descritta con riferimento alle accluse figure in cui:

la fig. 1 rappresenta una vista in pianta dello stampo, senza morsetti di tenuta, e

la fig. 2 rappresenta una sezione dello stampo lungo la linea II-II.

Con riferimento ai disegni lo stampo risulta costituito da una lastra 1 fig. 2 di polimetacrilato di metile, del colore voluto, opportunamente adagiata, tramite uno strato di carta di separazione 2, su una piastra di ferro 3, rigida ed indeformabile alla temperatura intorno ai 50° C. Lo strato di carte 2 viene fatto aderire alla lastra 1 di polimetacrilato di metile mediante una soluzione al 5% di gelatina animale.

La superficie superiore dello stampo è costituita, come di consueto da una lastra di vetro 4, che risulta distanziata dalla lastra 1 mediante i distanziatori 5 disposti tutto attorno allo stampo. Naturalmente lo spessore degli elementi distanziatori 5, che costituiscono in effetti granparte, nel senso dello pessore della superficie laterale continua dello stampo, è adeguato allo spessore della seconda parte della lastra bicolore che si vuole otte-

1

The first technology used by the Guzzini family was traditional and simple: heat to soften material slabs to make them adhere to a mould, which would give the product its shape. The process, known as thermoforming, is carried out in two stages: the heating of the material, then pressing – mechanically or by pressurised air – to achieve optimal contact with the mould.

In accordance with the local custom, horn was the material used, then replaced after the war with an artificial plastic material: methacrylate. Processing also moved on from the traditional manual method to simple wood presses, then later to completely automatic systems. Finishing operations were gradually reduced; thanks to the on-going refinement of plastic materials and moulding technology, it is today possible to manufacture a product that virtually requires no finishing at all.

Production went from a few articles crafted in horn of the early days, to Plexiglas utensils and containers by Fratelli Guzzini in the fifties, then subsequently to semi-spherical lamps by Harvey DH in the sixties, to Teuco showers and tubs of the seventies. The type of objects produced, their size and constituent materials have all changed – but the basic technological process implemented is substantially the same, even though thermoforming has been greatly facilitated by the use of vacuum, while computer-controlled equipment offers superior production performance.

Of all the innovations related to the industrial production of bathtubs, one of the most important was introduced by Virgilio Guzzini, who developed a moulding process on a vertical rather than a horizontal plane, which at the time was the convention. Production efficiency was thus increased, enhanced also by the gradual heating of material by means of methane combustion (instead of electrical resistances). In this way the structural characteristics of the plastic remained even throughout.

The cinema industry was also an important stimulus for technological development; in the movie "Barbarella" (1967) by Roger Vadim, the helmets, platforms and the transparent tub in which the star Jane Fonda bathed were manufactured by Polivar, the company incorporated in Rome by Gaetano Clementi and Virgilio Guzzini.

Gaetano Clementi, a graduate in Chemistry in 1946, after having followed the events of the war had developed an interest in the ability of aeroplane cowling to resist growing speeds and therefore strain.

2a

2b

1. The patent deposited on December 3rd 1958 by Fratelli Guzzini for the polymerisation of methacrylate multiple-layer two-coloured sheets.

2. a, b. These photos taken by Gianni Berengo Gardin depict the production of tubs by plastic sheets using the thermoforming procedure.

3. Three-compartment bowl (Giovanni Guzzini, 1955), a variation on the standard low bowl.

3

4

He thus discovered the properties of acrylic polymeric materials (used to make the cowlings), and decided to go back to the monomer, to exploit its resistance to abrasion and shock by other production techniques, without being bound by semi-finished sheets. The sheeting, produced during that period by Montecatini of Porto Marghera, were cut into pieces to make the plane cowling.
By using processing reject material, Clementi succeeded in splitting the polymer. He created a small workshop in his own home, where he began experimenting without following any known technological method, producing small objects, such as buttons and umbrella handles.
In 1950, during his search for factories that could supply him with polymethylmethacrylate rejects to obtain his monomer, Clementi became acquainted with the Guzzinis, who had used the material ever since 1938 to manufacture cutlery handles and other articles.
Realising the importance of this new development, the Guzzinis purchased the relative technology. In this way the system that Clementi had created within the confines of his cellar and garage was transferred as it was to the Guzzini factory, then extended.
Fratelli Guzzini had the practice of using two-colour sheets achieved by gluing, but this solution was messy, costly and above all, prone to delamination and detachment. It was, however, the plant manager Virgilio Guzzini who came up with the answer, obtaining a two-colour product by direct casting with monomers. This more economical system was finally patented in 1958.
The results can be seen in the numerous series of vibrantly-coloured containers with white interiors, which brightened the tables of many an Italian family until the early seventies.
Virgilio subsequently used his expertise of the new technology to pursue his research studies at the Polivar of Pomezia, founded by Clementi and the Guzzinis in 1959.
Clementi had in the meantime also developed a system for centrifuging acrylic monomers to produce tubes and then cylindrical objects, which in the early seventies were used to craft, amongst other products, *flutes* comparable to their glass counterparts, in three parts: the vessel, the decorated stem and the foot.
Large-sheet production and crafting were later perfected and refined, culminating in the new technology for the manufacture of bathtubs – thus laying the foundations of Teuco, incorporated by Virgilio and his brothers in 1970.

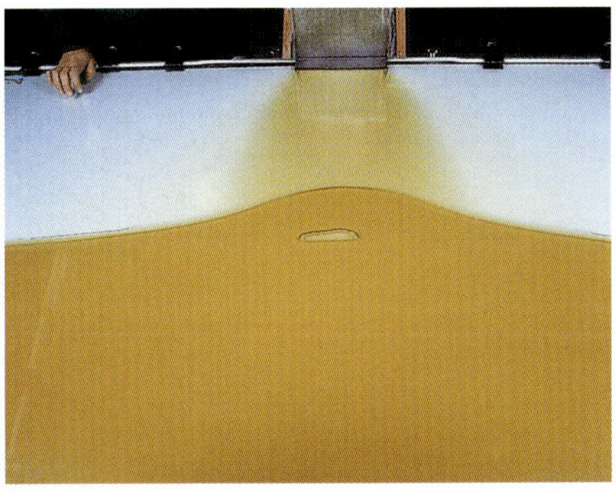

5a

5b

5c

5d

5e

5f

4. The *Bud* lamp (iGuzzini Studio, 1968) was made by blowing technology to form two-colour methacrylate, offering greater creative potential without the limitations imposed by thermoforming moulds.

5. Manufacture of a two-colour methacrylate product: (a, b) preparation of a twin-layer sheet; (c) moulding by thermoforming; (d, e) edge finishing and interior smoothing; (f) perforation of the finished product, a shade for suspended lamp.

1

2

Fratelli Guzzini was the first company in the Marche Region to realise the vital role played by design in conquering the market, as became clear when they would announce, "We want a designer to help us grow". The fact the designer was called Joosten or Capici meant quite simply, "I want a name that gives prestige to production".

Most of the articles initially produced were of the conventional type, both in terms of the items themselves as well as their furnishing role, while a small portion of household objects featured Swedish, Danish and even American ranges. Italian-design products were solely distributed through certain select shops that tailored for the discerning customer, accounting for some 5% of the market. This was the situation through to the sixties.

What were the defining characteristics of these new advanced production lines? Clearly, very few links with the past. The presence of wood in the Nordic products was effectively a link with tradition, but by the same token, that kind of produce was emblematic of a rational, Calvinist lifestyle. This kind of product differentiation was a distinctive market feature. The difference as a competitive edge.

Guzzini needed to give greater substance to plastic material, to the moulding sheet, fully exploit its thickness, its transparency, its opacity. Plastic was indeed not conceived as something lightweight or easily perishable. Research was necessary not only on the material itself, but also on the context of the object: a concentration of attention on the kitchen habitat.

At this stage Luigi Massoni, who at that time was in the employ of Boffi, made his entry to the company. "My activity at Boffi gave me the opportunity to follow the developments in furniture items, and in particular of what is transferred from the kitchen to the table. In 1965, a kitchen that I designed for Boffi was displayed during the Batimat fair in Paris; the kitchen was characterised by its modularity, to a basic measurement of 15 cm, horizontally and vertically. This innovation actually defined the optimal measurement for ergonomic performance, later officially adopted by the UNI.

There was no need to conduct similar studies on object dimensions at Fratelli Guzzini, but what is clear is that the objective I tried to attain for Boffi was in a certain way transferred to Guzzini. Objects can't be designed without accounting for their ergonomic, therefore their dimensional, features."

This new cultural approach introduced by Massoni was one of his most original contributions. He carried out his work in line with his research on the development of artisan skills, on using material to exploit its full technological potential, as well as that of the context and the determination to bring Italian design to serial production.

"It is important to recall that we worked as a team; indeed one of my qualities is that I design in tandem with the producer. This was certainly one of the keys to success, because it helped technicians and workers become more appreciative of the material, better acquainted with its enormous potential. Our designs were completely stripped of the superfluous to highlight the dimensions, the material, the finish. The approach was no longer one of covering the surface with decorations, but rather of showing the mate-

3

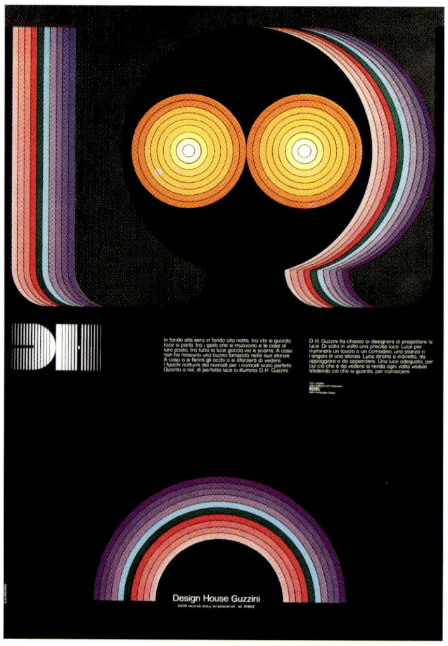

4

1. *Cespuglio* table lamp designed by Ennio Lucini in 1969 as a tribute to designer Gino Marotta.

2. The *Brumbry* table lamp (1968) was made up of a white or coloured methacrylate shade set on a chromed metal base. The shade became particularly bright due to the four incandescent bulbs.

3. *Joint* chair designed by Luigi Massoni in 1972.

4. Poster designed by Mimmo Castellano in 1970 for personalisation of iGuzzini sales outlets.

5

6

rial for what it was, exalted by the use of colour. In fact we did a lot of colour tests: a satin-finish for black, using yellows, reds and oranges – colours that made the kitchen and the table more joyful and warm, banishing the sense of coldness that we still carried inside with our memories of the war. We strived for a touch of festivity, which was appropriately batched and transferred to these objects. Graphic artists were brought in to express the tastes of that period, which had evolved from the previous era. I had some of these artists, such as Bob Noorda, decorate the surface of these objects, abandoning the classical patterns". Distribution by large department stores – the most important was la Rinascente – helped draw attention to the issue of the visibility of the product and its packaging, and this in turn had an impact on subsequent research. By means of individual products, Guzzini offered a new way of construing the table through the objects that cross it.

Massoni's contribution went beyond kitchenware to light fittings – but unlike the case of Fratelli Guzzini, he had the opportunity of working in a freshly-constituted company, without having to account to consolidated products or images, and this was a great stimulus to innovation. Other designers became involved: Carlo Emanuele Ponzio, Gio Ponti, Cesare Casati.

"We built whatever was necessary to be able to draw other designers to us. Everything that accompanied the object had to be created: the printed paper that showed it, its launch, advertising, the catalogue, graphic designs, starting with the product logo. When I needed help, I called in some designers from Milan; Ennio Lucini was the first, followed by Mimmo Castellano".

After the initial period, the company began to become known throughout the Italian market, and even abroad.

"At one stage, in 1966-67, the need arose to form DH, because distribution varied, with stores more attentive to aesthetic aspects, with higher-priced products. This opportunity to expand our turnover required that we come into line with a clientele who wanted to live in a home different from the conventional, who had the money to aspire to a higher standard of living. Various furnishing companies had already oriented their production towards this model, becoming in fact standards of reference. This 10 percent of non-traditional consumers generated an enormous quantity of business.

In lighting we probably moved more efficiently and rapidly, where we'd found a more fertile terrain. In household products, a market already existed so what was demanded hardly ever coincided with what I could design, while for iGuzzini we came up with new designs". iGuzzini illuminazione decided to opt for a twin distribution channel: specialised and furnishing shops were served by DH, and wholesale outlets by Harvey; this suited a broader section of the community thanks to the lower prices offered.

At one stage Adolfo Guzzini realised that all production types were being directed towards retail outlets, so he focused on lighting technology, strengthening an industrial approach with Massoni. "You were part of the company directive staff, in the truest sense of the word. There was total understanding between designer engineers, producers and advertising".

5. *Jolly* bed tray, designed by Luigi Massoni in 1970.

6. *Piccola and Media* table lamps, designed by Gio Ponti in 1967.

7. The traditional fondue plate (1971), besides being reinterpreted in a plastic version, was behind the party plate designed by Luigi Massoni (1972) to reflect the changing, more informal altitudes to eating.

8. During the sixties, the shape of products was often emphasised by patterns; in this case, the *Ornella*, by Ornella Noorda (1968), applied in various versions not only to bowls, but also trays and cutlery.

7

8

1

1. The processing of polycarbonate globes.

2. The *Multiplo* designed by Rodolfo Bonetto in 1970.

3. Modular opaline methacrylate set (1969) featuring printed decorations by Ornella Noorda.

4. Soft drink and water glasses, *Riviera* (1964).

Parallel to the developments in organisation and economic scenarios, the company also witnessed substantial changes to products and the way they were manufactured, by the introduction of new technology and materials. Between 1960 and 1970, formal technical experiments helped assure the production of innovative articles over the next decade. Pipe moulding tests were for example conducted for the production of carafes and glasses. At the same time, the strategy of concentrating on a small number of wide-distribution products was defined and consolidated. Objects were decorated by silk-screen printing, with designs for ensuring production economy as well as formal, functional qualities. During this decade of research, the foundations on which the fortunate collection of *Chef line* tableware objects, designed by Bruno Gecchelin between 1980 and 1985, were consolidated. In the lighting industry, the range of materials was extended by new technical solutions, such as by the use of ABS, a relatively inexpensive plastic with remarkable aesthetic properties, subsequently seen in the *Multiplo* by Bonetto (1970). The possibility of using metal bodies was also explored, a successful example of which was the Lucciola halogen lamp (1972) by Fabio Lenci. The production of spotlights was indeed to become one of the most important lines, the product type satisfying the new domestic and commercial lighting demands of modern interior architecture, in the home as much as in sales outlets. To manufacture light fixtures economically, decisively industrial technologies were adopted (such as drawing), thus definitively shifting the Guzzini production identity from the workshop towards that of the factory.

An important step forward was taken in plastic material technologies with the introduction of injection blowing, a technology for the production of hollow volumes. It was Giannunzio who returned from a convention with the news of a new technology designed for small volumes, realising subsequently its potential application to larger ones. The eventual result was a Guzzini world patent for the production of spherical acrylic and polycarbonate bodies, a veritable milestone in company history, as seen in the development of some of the most fortunate designs of the decade. A technology that not only assured large profit margins, but that also contributed to product diversification, thus rationalising the entry into the sector of lighting fixtures.

2

3

4

5

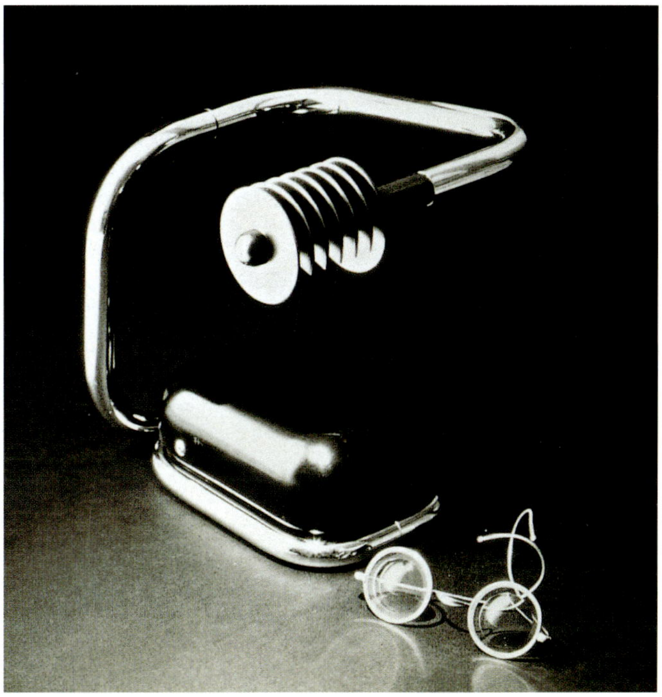

6

7

8

5. Advertisement (1965) featuring photograph of the suspended *Medusa* by Massoni.

6. *Lucciola* bedside table designed by Fabio Lenci, 1972.

7. *Aquarius* bathroom fittings and tub by Fabio Lenci, 1972.

8. *VIP* set with smoking and desk accessories.

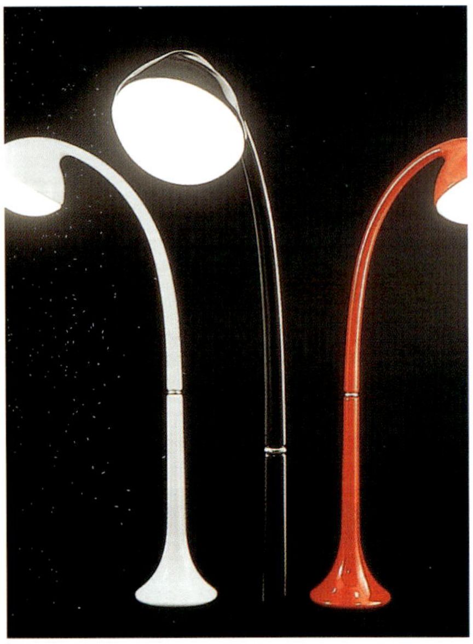

1a

Fabio Lenci is the designer who has made the most significant contribution to the structure and production of Teuco.
Lenci became acquainted with the Guzzini brothers in 1963-64, when he proposed a number of projects from his furniture store in Rome.
One year later the company – which at the time was called Harvey DH Design House – asked him to design some lamps. This opportunity allowed Lenci to develop his acquaintance with the Guzzini brothers, their culture and mentality for which he felt a natural affinity.
Harvey DH was a small enterprise that employed between twenty and thirty people; thanks, however, to the contribution of Massoni, it had a clear idea of the importance of research to gain a competitive edge, as witnessed by the relatively large size of the engineering department in relation to the overall dimensions of the company. In this context, Lenci initially endeavoured to improve existing product types, then subsequently to introduce authentic design innovations, in particular domestic light fixtures. Lenci collaborated in the production of lamps for some seven years, until the company turned its attention to lighting engineering.
From the technological viewpoint, Lenci started out with the material most commonly used by the company – blown acrylic – with which the Guzzinis produced the design known as "three quarters of a sphere". On the basis of this idea, Lenci designed support spheres, such as the *Focus* lamp (1967), featuring a wire support. He then turned to the development of new types of light sources, such as quartz and iodine lamps that he had discovered in slide projectors, in which they were used due to their colour temperature similar to that of sunlight. Lenci decided to use these lamps for lighting fixtures, starting to work on halogen lamps (which due to their low voltage required a transformer). The *Lucciola* (1972) was the first halogen lamp model placed by the company on the market.
He subsequently conducted experiments in the production of polyurethane light fixtures while also carrying out research with Bayer for the use of rigid polyurethanes, such as Baydur. This material, that Lenci had already used to produce tables, proved to be also suitable for outdoor light fixtures.
"I was still a consultant for iGuzzini illuminazione", recalls Lenci. "I worked for them during the day at Polivar, where I met Virgilio Guzzini in the evening, after my day in the engineering office, then worked through to

1 a. *Lampione* outdoor lamp, the first to be made by the new technology using rigid polyurethane (Fabio Lenci, DH Guzzini, 1971).
b. Sketches by Fabio Lenci for the *Lampione* (1971).

2. Sketches by Fabio Lenci for the *Focus* (1969).

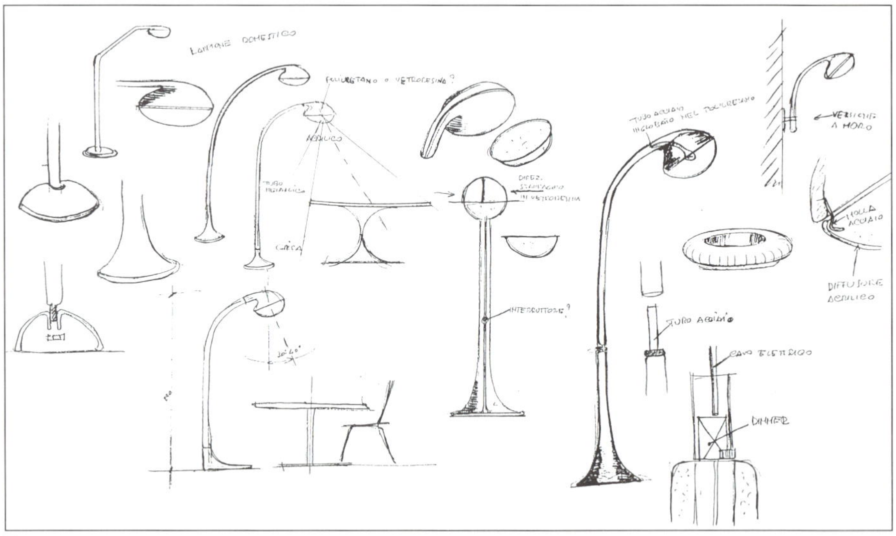

1b

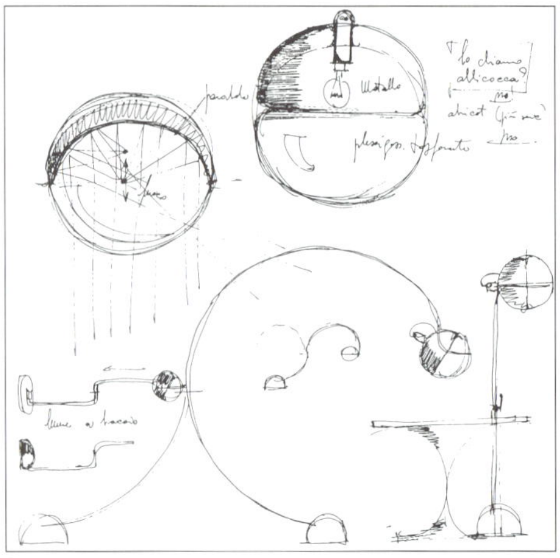

2

midnight or even one, in a kind of warehouse where they pressed large acrylic sheets". Virgilio Guzzini decided to create a new company for these projects: Teuco. After a few years of relative mistrust shown by the market – which only knew ceramics and steel – and in particular after the round shower, Teuco products began to become increasingly successful.
One year after the incorporation of the company, Teuco was joined by Antonio Renzi, then partner and sales manager, with whom Lenci immediately felt in harmony as regards the form of products to be presented to a market that was beginning to associate the bathtub with the idea of comfort and relaxation.
It was during this time that Lenci turned to the design of a sauna, the feasibility of which was tested in the designer's own study. "I obtained a round shower, heated it with a set of hairdryers, then sealed the top – but it produced so much heat that the shower almost buckled. I then noticed that at a lower temperature a Turkish bath could be created. The idea remained at the experimental stage for some time, until Antonio Renzi found in the USA a device which, according to my specifications, allowed the inclusion of a Turkish bath function in the shower". Following the advent of electronic technology to control electrical mechanisms and pumps, the Turkish bath asserted its role, even over that of the shower. The innovation of habitual product types is the hallmark of the projects developed by Lenci with the assistance of Giovanna Talocci. After the round shower and the sauna-Turkish bath, a number of other new products were designed with a whirlpool, such as the 2 x 2 m tub-shower for two people, of which at least one thousand were sold.
Another example of innovation was the tub with a glass panel. "In this case I started out from a reflection on human desires, including that of being able to see underwater: a window allows the body to be seen from outside the tub". Originally produced as a prototype for an Australian exhibition dedicated to innovation in Italian design, the tub was later produced from 100% acrylic, featuring a panel with soft upper section that doubles as a seat and panel frame.
Teuco has anticipated other market trends by original experiments in the field of ultrasounds, an initiative that remained top secret for three years, during which Lenci and Talocci designed shapes for ultrasounds most suited to the particular product model.

3

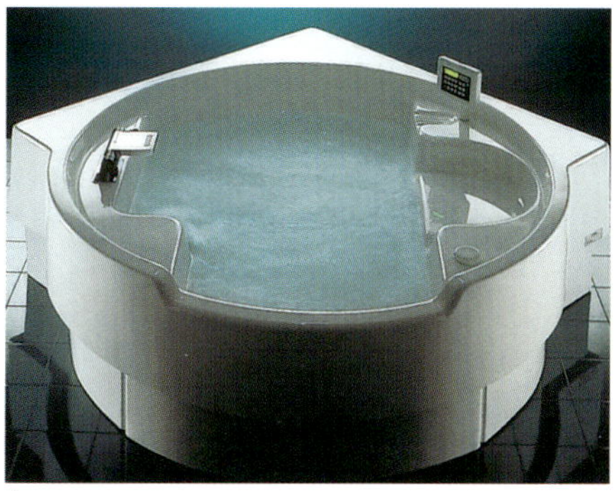
4

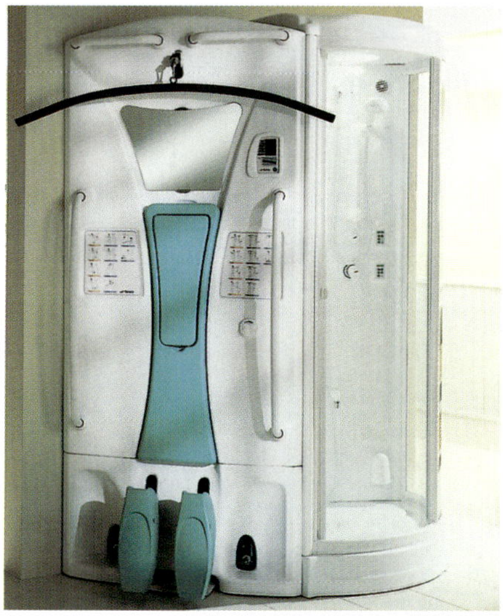
5

6

3. Corner hydrosteam unit with equipped wall featuring water massages (Lenci and Talocci, Teuco, 1993).

4. Tub for two, with whirlpool, designed by Lenci and Talocci (1975).

5. P01 *Fitness Corner*, also featuring shower (Studio Lenci, 1999).

6. Range of coloured sinks designed by Carlo Urbinati and produced by Teuco (1974).

"They looked me up in Milan, where we met in a hotel. It was probably in 1976. I always started out like that with everyone: I was sought according to the success that a previous product had achieved". This is how Bruno Gecchelin, designer, recalls his meeting with Adolfo Guzzini, who asked him to design something truly original for iGuzzini illuminazione, something that did not simply rehash past experiences.
At that time the company was undergoing a transformation from a classical-modern lighting company into a lighting engineering enterprise. Modern-lighting companies were rapidly growing in a context that was ideal for Milan, but not for Recanati; lighting engineering, however, was securely in the hands of German or US companies. Gecchelin began his collaboration with Adolfo Guzzini to design lamps with a specific engineering performance, thus giving rise to a brand-new corporate strategy.
The first model was the *Praxis*, an operative table lamp based on a characteristic patented friction system. This was followed by the *Vela*, a decorative wall luminaire made entirely from plastic, exploiting the traditional Guzzini skills in crafting this material.
Other products included the *Ialea*, a tiny table lamp made of two identical shells held together by a ring and with central socket; and the Open system, the first in a fortunate series of lamps that continues to this day. The watershed was, however, the *Virgola* (1979). At that time, halogen lamps were an exciting innovation, in particular cylindrical halogen lamps and twin couplings. The design played on the compactness of the object and the option of making the object in two parts, shell against shell, assembled by two screws. The optics were calculated to guarantee a specific degree of lighting efficiency. These features made the model especially competitive compared to conventional halogen lamps.
Virgola effectively anticipated the formal and technical characteristics that would reappear in other models: the bevel at 45°, miniaturisation, die casting. It also hailed in a new approach to operation by the company: casting was contracted out to third parties. For the first time, capital was being invested in outsourced moulds that would return a product that was almost finished, a totally new experience for the company. In exchange for the new running costs, it benefited from obtaining high production quantities that were impossible to achieve inside the plant. "As regards the moulds", recollects Gecchelin, "outsourced activities were very efficient. The real prob-

1a

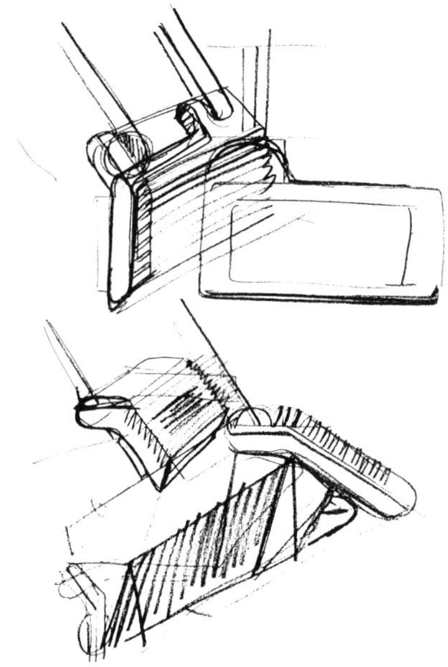
1b

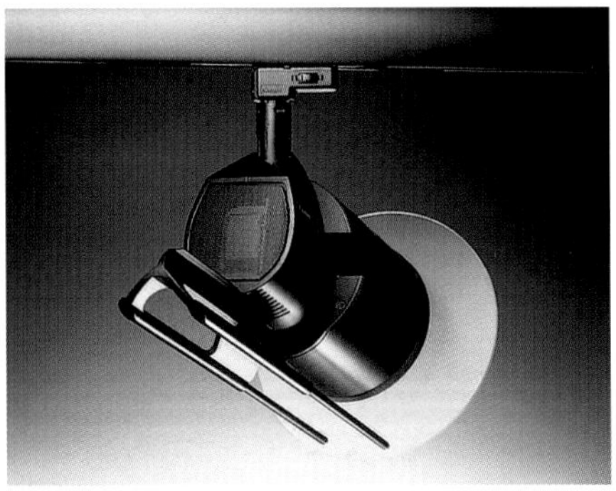
2a

2b

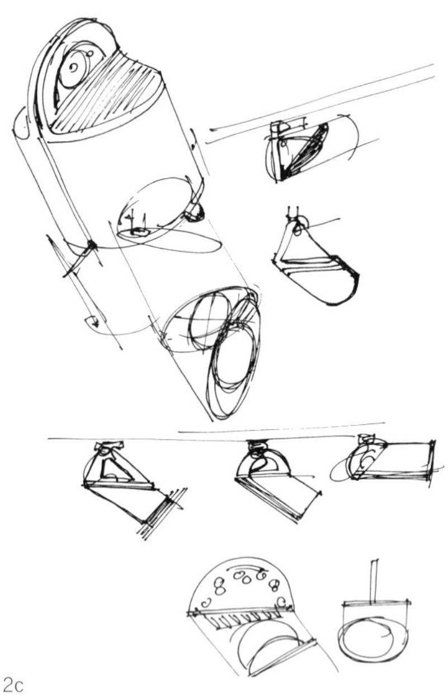
2c

1 a. *Virgola* halogen lamp lighting system. One of the first systems to be designed with "bivalve" die casting technology. Safety is assured by the fireproof glass (Bruno Gecchelin, iGuzzini illuminazione, 1979).
b. Sketches by Bruno Gecchelin for the *Virgola*, 1979.

2 a, b. *Shuttle* spotlight system, made from die-cast aluminium, can fit 15 different types of low and mains voltage bulbs. Flux intensity is adjusted by a dimmer. The *Shuttle* was awarded the 15th Compasso d'Oro prize, 1989 and iF awards, Design Auswaahl, Oscar Snai (Bruno Gecchelin, iGuzzini illuminazione, 1993).
c. Sketches by Bruno Gecchelin for the *Shuttle* system.

3. Bruno Gecchelin, *Chef Line* cutlery drainer.

4 a. The Chef Line kitchen utensils was designed to bring the benefits of professional equipment to the domestic setting (Bruno Gecchelin, Fratelli Guzzini, 1981- 84).
b, c. Sketches for *Chef Line* utensils.

3

4b

4c

lem was how to use them. In the company we needed people who were able to control the transformation, the growth of an industry specialised in technologically-advanced products". Painting was in fact the only activity to remain in-house.
Commercial success came immediately, and Gecchelin was asked to continue along the same line. The problem was the planning of the product portfolio for a market that was potentially receptive of each product. What comes first? The neon, or outdoor luminaires? Recessed fixtures or another lamp type? "The ideal thing would have been to spend time constantly in the company, as did Massoni. But an artificial situation came about, like when someone comes to visit at home, you tidy everything up and everything works well. After which, when I had gone, it all turned back into the province. I think that it was then that an active atmosphere should have been created, but the company wasn't ready. Commercial success made things grow, but almost automatically".
The *Virgola* was followed by other products, the most important of which was Prisma, a fluorescent system that answered the need to use neon tubes simply, created in 1984.
Of the more recent experiments in light sources, the *Shuttle* series was created in 1987, comprising all lamp types. It was *Shuttle* that brought iGuzzini illuminazione into the international limelight when it was awarded the 15th Compasso d'Oro ADI in 1989.
After approximately a year of working with iGuzzini illuminazione, Gecchelin made the acquaintance of Giovanni Guzzini, who took him to visit the household articles factory. On behalf of Fratelli Guzzini, Gecchelin designed amidst other things the *Chef Line* series of utensils and kitchen containers, with special attention to the needs of professional objects. Following his initial sketches, Gecchelin took his technical drawings to the company, going beyond the consolidated practice by which the designer (or Giovanni Guzzini himself) worked in close contact with the modeller to create a scale model for testing purposes. From this viewpoint Gecchelin contributed by offering Guzzini companies the opportunities to stimulate the internal evolution process, to open up to the exterior with greater competence.

4a

If today it is virtually impossible to see the night sky in the city, the fault lies with poor lighting standards. The incorrect design of lighting systems leads to the dispersal of luminous beams, glare and reflection in undesired directions. The problem concerns not only astronomers and environmentalists, but above all professionals who have to deal with technical lighting: architects, designers, public administrators, corporate managers.
Action must therefore be taken along two lines: on the one hand, by raising public awareness, and on the other, by training professional operators to allow them to find even more efficient solutions to the problem.
The iGuzzini illuminazione awareness-raising campaign in 1982 made reference to Goethe's invocation for "more light" to introduce the themes and problems of lighting. Numerous other campaigns then followed, with pages in the most important national newspapers and magazines. Attention was drawn to various practical issues: light as a deterrent for small-scale crime, energy saving strategies, the physiological effects of light. Interior light-

1a

1b

1c

1d

1. Three designs for the campaign against light pollution: (a) a newspaper advertisement, 1997; (b) the cover of an ilibriGuzzini collection, published by Editoriale Domus and edited by the Centro Studi e Ricerca iGuzzini, 1998; (c, d) two campaign inserts, designed and produced by the STZ agency, Milan (c. 1994), (d. 1997).

ing was also covered, at the home and at work (correct lighting as a remedy to fluorescent flickering and annoying computer monitor glare). It was also shown how the appreciation of monuments and other cultural assets is directly linked to correct illumination.
In response to these problems, iGuzzini offered a number of products designed *ad hoc*, which eventually led to different series of luminaires applicable to a range of similar situations. Some examples of this strategy include the projects for the Turin Lingotto, the Galleria Borghese in Rome and the Centre Georges Pompidou in Paris.
On the professional side, iGuzzini provided designers with all the experience accumulated over the years, including the facilities of the Centro Studi e Ricerca. As of 1995, it has promoted general training activities, research programs and conservation of the corporate historical heritage. Its activities also include the publication of volumes for professional users that illustrate the state of the art in lighting engineering research.

2

3

4

2. The Château de Méry, Paris. Design by Jean-Michel Wilmotte, 2000.

3. Illumination of sculpted works in the Museo Borghese, Rome, 1997.

4. The Galleria Vittorio Emanuele in Milan. Project designed by Piero Castiglioni, 2001.

1. The *Cestello* system, designed by Gae Aulenti and Piero Castiglioni (1994), given the iF Design Award by the Industrie Forum Design of Hannover in 1997.

The complex intertwining of research and production, organisation and technological development, market and ideas has settled over the years to create a professional culture that has proven to be a key strength and source of identity of Guzzini group companies. An identity that has been explicitly acknowledged in the motivation of a number of prizes awarded to the group as a whole. Of particular importance, the ADI Compasso d'Oro received in 1991, awarded "for having developed over the years a coherent production and design ethos, in which design culture has been the common denominator and distinctive element". A number of iGuzzini illuminazione products have been awarded prizes for overall excellence, beyond their individual features: such was the case of the Industrie Forum Product Award awarded in 1997 to five lighting systems: *Kriss and Base Lighting* (design by Yamada), 1991 and 1993; *Woody* and *Miniwoody* (design by Mario Cucinella), 1996; *Cestello* (design by Gae Aulenti and Piero Castiglioni), 1994; Nuvola (design by Renzo Piano Design Workshop), 1996.
Awards for individual products include the ADI Compasso d'Oro for the *Shuttle* lighting system (designed by Bruno Gecchelin) in 1989 and the *Nuvola* outdoor luminaire (Renzo Piano Design Workshop) in 1998.

During the early seventies, the round shower was exhibited at the MoMA in New York, then placed on permanent display at the Museum of Modern Art of Philadelphia. In 1988, the famous corner bathtub was selected to represent Italian technology in Brisbane, Australia.

1

2

2. *Nuvola* outdoor lighting system (1998) by Renzo Piano Design Workshop, awarded the 18th ADI Compasso d'Oro.

3

4

5

6

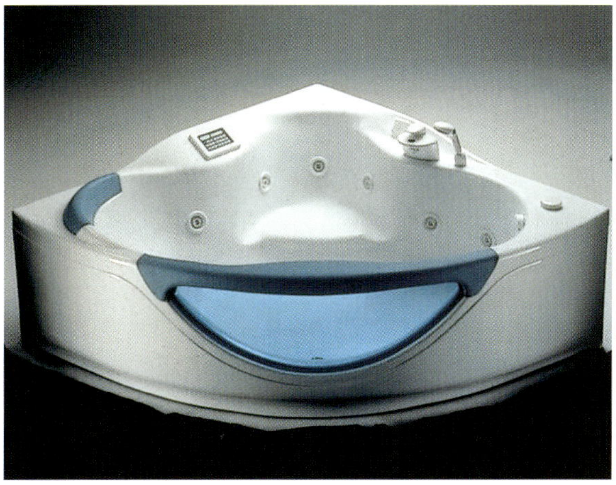
7

3. *Isotta* pasta pot designed by Bruno Gecchelin. The design received an honourable mention at the 18th Compasso d'Oro.

4. *Portaspago* designed by Bruno Gecchelin. Recipient of an honourable mention at the 14th Compasso d'Oro.

5. *Ala* trays designed by Bruno Gecchelin, honourable mention at Design Plus.

6. The first round shower cabinet designed by Fabio Lenci in 1972.

7. Corner bathtub with whirlpool 266. Made from polyurethane and acrylic, it features seat and headrest, as well as coloured window and underwater light (Fabio Lenci and Giovanna Talocci, Teuco, 1994).

Documents

The Enrico Guzzini family tree

Enrico Guzzini
1862-1948
Annunziata Frapiccini

- **Ubaldo** 1894-1984 — Rosa Giordanelli
  - **Lidia**, **Enrico**
- **Pierino** 1899-1981 — Alessandra Marucci
  - **Vinicio**, **Maria**, **Alberto**, **Ines**, **Armida**, **Cherubino**, **Ada**, **Igino**, **Elsa**, **Lucia**
- **Celestina** 1902-1995 — Giulio Moroni
  - **Giuliana**, **Giuliano**, **Laura**, **Maria**
- **Mariano** 1904-1986 — Irene Guzzini
  - **Giovanni** 1927 — M.Pia Miccini
    - ***Paolo*** 1953
    - ***Marina*** 1958
    - ***Gianluca*** 1961
  - **Raimondo** 1928-1978 — Giuseppa Galdi
    - ***Mariano*** 1956
    - ***Domenico*** 1959
    - ***Emma*** 1962
  - **Virgilio** 1929 — Edy Rafanelli
    - ***Marco*** 1958
    - ***Mauro*** 1961
  - **Marcella** 1932 — Giorgio Giorgetti
    - ***Amedeo*** 1956
    - ***Andrea*** 1957
    - ***Mariano*** 1962
  - **Giuseppe** 1935 — Silvana Sabbatini
    - ***Gabriela*** 1968
    - ***Carlo*** 1970
    - ***Rossella*** 1974
    - ***Alessandro*** 1976
  - **Giannunzio** 1938 — Maria Mattutini
    - ***Monica*** 1963
    - ***Francesco*** 1968-1987
  - **Adolfo** 1941 — Anna Taccaliti
    - ***Massimiliano*** 1969
    - ***Matteo*** 1972
- **Maria** 1905 — Antenore Corvatta
  - **Enrico**, **Annunziata**, **Beniamino**
- **Silvio** 1913-1984 — Marzia Sabbatini
  - **Paola**, **Serenella**, **Stefano**

The family and the companies: a chronology

The over 85 years of business activity by Guzzini companies have involved the participation of four generations of entrepreneurs and workers.
In 1912 Enrico Guzzini, with the assistance of his family, started up a small activity for the production of hand-made horn snuff boxes.
As of the twenties, Enrico's elder sons – first Pierino, then Mariano and later Silvio, the youngest – introduced mechanised and electrically-powered devices to the production process.
Pierino decided to focus on modernising the family business, while Enrico retired, handing down management entirely to his sons. The year 1938 saw the Guzzinis introduce the use of Plexiglas sheets in the production of household articles, a development that opened prospects previously unimaginable.
The three brothers decided to divide company roles between them. Pierino took charge of administration, as well as sales and technical management. He was also involved in certain important technological innovations, such as moulding with compressed air. Mariano assumed responsibility for management of production activities, product quantity and quality standards, while tending relations with employees and suppliers. He created new productions in collaboration with Pierino, and also instructed methods and techniques to other workers. The youngest son Silvio was the expert in horn crafting, which he continued to look after personally.
All three sons were jointly and severally administrators of the small company, each with equal powers.
With the advent of the war, many young workers and employers, including Silvio Guzzini, were called to arms. In 1941, Pierino's and Mariano's elder sons, hardly 15 years old, were forced to leave school and enter the factory to learn the trade and help their parents, assuming different duties: Vinicio took charge of production planning, while Giovanni joined his father in the technical-production sector (construction of moulds, prototypes and new products). Thus the elders of the younger generation entered the company, later followed by their brothers and sisters. The post-war forties saw a gradual resumption in activities and employment of workers.
By the fifties, Pierino had 10 children, Mariano 7 and Silvio 3. They represented a vital entrepreneurial resource: young boys and girls, having reached the working age, learn the activities in each company area, in different departments, with dedication and profit. Some later turned to other study or work activities, or established their own families; others (still present) remain with the company.
In 1953, the 24-year old Raimondo set out to travel throughout Italy to relaunch sales, to become better acquainted with the market and the clientele.
In 1954, following the purchase of a chemical formula and process, the Guzzinis created on licence an exclusive chemical plant for the production of two-colour cast acrylic sheets.
The plant, constructed with the assistance of Gaetano Clementi and Pierino Guzzini, was run by Virgilio Guzzini, who became an expert in this type of production process. The technology thus developed gave rise to the two-colour sheets used in tableware product collections, characterised by their innovative crafting and design. It was at this stage in time that design became part of a strategic approach developed jointly with important professional designers.
In 1955 the company purchased its first injection press for the production of glasses and cutlery with acrylic granules. The new technology, which paved the way for the restructuring of the sixties, was entrusted to Alberto Guzzini.
Other young family members gained their first experience in different areas of the company: Cherubino, firstly in production, then in sales; Giuseppe, who joined Pierino in administration and management; Giannunzio and Igino, both involved in the technical-production area.
The children of the founding partners decide to direct and develop the company towards new directions, roles and responsibilities; the expanding markets were in fact fully receptive to the prospects of greater production, though product diversification was necessary. Intense meetings and feedback between the two generations gave rise to new strategic lines for all: as of 1960, the heirs of the three founding members opened new businesses not in competition with one another, nor with Fratelli Guzzini.

The new enterprises include iGuzzini, Acrilux and Teuco.

During these events, each of the main companies incorporated in turn a number of subsidiary, hi-tech companies to serve the parent company, generating even greater know-how for the Group as a whole.
Today's Guzzini companies, all located around Recanati, employ no less than 1,631 workers, representing a resource for the apprenticeship and training of members of the fourth generation who, their studies completed, entered the Group companies.
Fratelli Guzzini:
Domenico (1987), Massimiliano (1999)
iGuzzini illuminazione:
Paolo (1978), Marco (1987), Alessandro (2001)
Teuco:
Mariano (1986), Monica (1987), Gianluca (1990), Mauro (1997), Carlo (2001).
The organisational development of the corporate group eventually led to the establishment of three holdings for management of personal shares of the three families of origin. The first was Fimag (Finanziaria Mariano Guzzini), created in 1982, allowing the grouping of all share packages of each enterprise into a single *new company*.
Fimag's role is to appoint the administrators to controlled companies, while dealing with problems arising in interfamily relations and safeguarding the managerial autonomy of each company.
Fimag also has the function of developing and approving general strategies of controlled companies, besides acting as a forum between elder and younger family members for discussion of problems and management strategies. It is also the centre for assessment and self-assessment of young family members.

Fratelli Guzzini production plant (2000).

Fratelli Guzzini
Fratelli Guzzini continues its operations in the household article sector. In 1967 the company underwent substantial reorganisation: director shareholders Pierino and Mariano transformed the company from a partnership into a private limited company, transferring their share quotas, then retiring.
The company thus passed into the hands of the third generation, guided by a board of directors chaired by Silvio Guzzini and two directors – Cherubino and Giuseppe Guzzini, to represent the other two families.
The new thrust given to organisation, not to mention the generational turnover, helped the parent company grow even further, as seen in the structuring of more defined functions and duties: sales and marketing under Raimondo Guzzini; production management under Giovanni, who had in the meantime acquired broad expertise in production processes, mould and prototype construction. In 1968, at the instigation of Raimondo and with the consent of all, Rede Guzzini was incorporated to include the 11 partners into the company structure.
In 1970, the burgeoning Fratelli Guzzini company inaugurated its new production site in San Leopardo di Recanati. The event represented an organisational, cultural and entrepreneurial milestone, an endeavour towards rationalisation as well as a step towards greater productivity. The investment was equal to 35% of yearly turnover (some 900 million lire). The most important consequence, however, was that the decision gave everyone involved, shareholders and workers, the tangible certainty of having finally become a modern industrial enterprise.
During the same period, Fratelli Guzzini was joined by Silvio's son Stefano, who had recently graduated in engineering. He became sales manager and company managing director, posts Stefano occupied through to 2001, when he relinquished his shares and left the company.
Fratelli Guzzini became a leading company in its industry; in 2000, it employed 300 workers and was present in no less than 60 countries throughout the world.

Central offices of iGuzzini illuminazione, designed by Mario Cucinella, 1996.

Central offices of Teuco in Montelupone, designed by Massimo Picaciamarra, 1996.

iGuzzini illuminazione

New product lines – such as decorative lamps – were developed for new markets in the late fifties by all of Mariano's children involved in the company.
In the same period Adolfo, having completed his studies, then Giannunzio, following his apprenticeship at Fratelli Guzzini, entered Harvey to start their careers with new responsibilities.
After an initial period of artisan-type activity, the first real stage of development took place over the decade 1963-74, when production efforts were concentrated on light fittings. Of greatest significance however were the new partnerships established with designers, an event that would change the course of iGuzzini illuminazione.
Another turning point was encountered in the mid-seventies. Competitive strategies, the backbone of the company, were elaborated: lighting engineering, technological innovation, internationalisation.
Lighting engineering is a discipline combining ergonomics and architecture, physiology of perception and design, psychology and technology – a discipline that demands great skills and productive flexibility, the capability of renewing materials and processes.
Invitations to cooperate were therefore extended to designers with extensive engineering know-how, who adopt design criteria able to satisfy the variety of lighting needs of each venue and application.
iGuzzini researchers were joined by doctors, sociologists and physicians from the most renowned international research centres, because design implies a deep awareness of all facets relating to light: Harvard University, Boston MIT, "La Sapienza" University of Rome, the Milan Politecnico, Istituto Centrale per il Restauro, the CNR.
The technical skills developed also drew the collaborative efforts of the most prestigious names in architecture, such as Renzo Piano, Richard Rogers, Norman Foster, Gae Aulenti, Michele De Lucchi, Odile Decq and Benoît Cornette, Rafael Moneo, Vittorio Gregotti, Kenzo Tange, Mario Botta.
In 1995 the Centro Studi e Ricerca was opened to contribute to the cultural debate, reflecting the company's conviction that the design of lighting projects plays a vital role in many spheres of human activity: work, culture, social relations.
In 2000, iGuzzini employed over 709 workers in eight branches throughout Europe.

Teuco

Virgilio Guzzini moved to Rome in 1960, to the new company opened by Gaetano Clementi, with the objective of developing high-tech applications through the production and processing of special cast sheets of polymethylmethacrylate (PMMA). The invaluable experience acquired there eventually led to the creation of Teuco in Recanati. It was in fact in 1972 that Virgilio together with his brothers and Clementi decided to start producing sanitary equipment from acrylic sheeting.
Recanati hosted the site in which the first products were manufactured: a round shower cabinet, then a complete range of bathroom accessories. In later years with Antonio Renzi, the first sales manager, then later partner and general director, the first whirlpool tubs were made. Teuco history began with the renewal in the perception of the bathroom – both as a setting and a technological concept – of which Teuco was an authentic protagonist, introducing innovative solutions and a totally-new aesthetic approach.
The first in the world to manufacture showers from thermoplastic material, Teuco specialised in thermoforming, a processing method that allows the creation of harmonious, ever-new shapes and contours with methacrylate sheeting. More recent years have seen the introduction of injection and coinjection processes, innovations that assure an even higher quality product, also in terms of durability, safety, function and design.
Due to the lack of available industrial land in the area of Recanati, in 1996 Teuco moved its production site and offices to Montelupone. With its 556 employees and branches in France, Germany, Great Britain and Spain, it is today the undisputed European market leader in whirlpool tubs.

Structure of the Fimag group

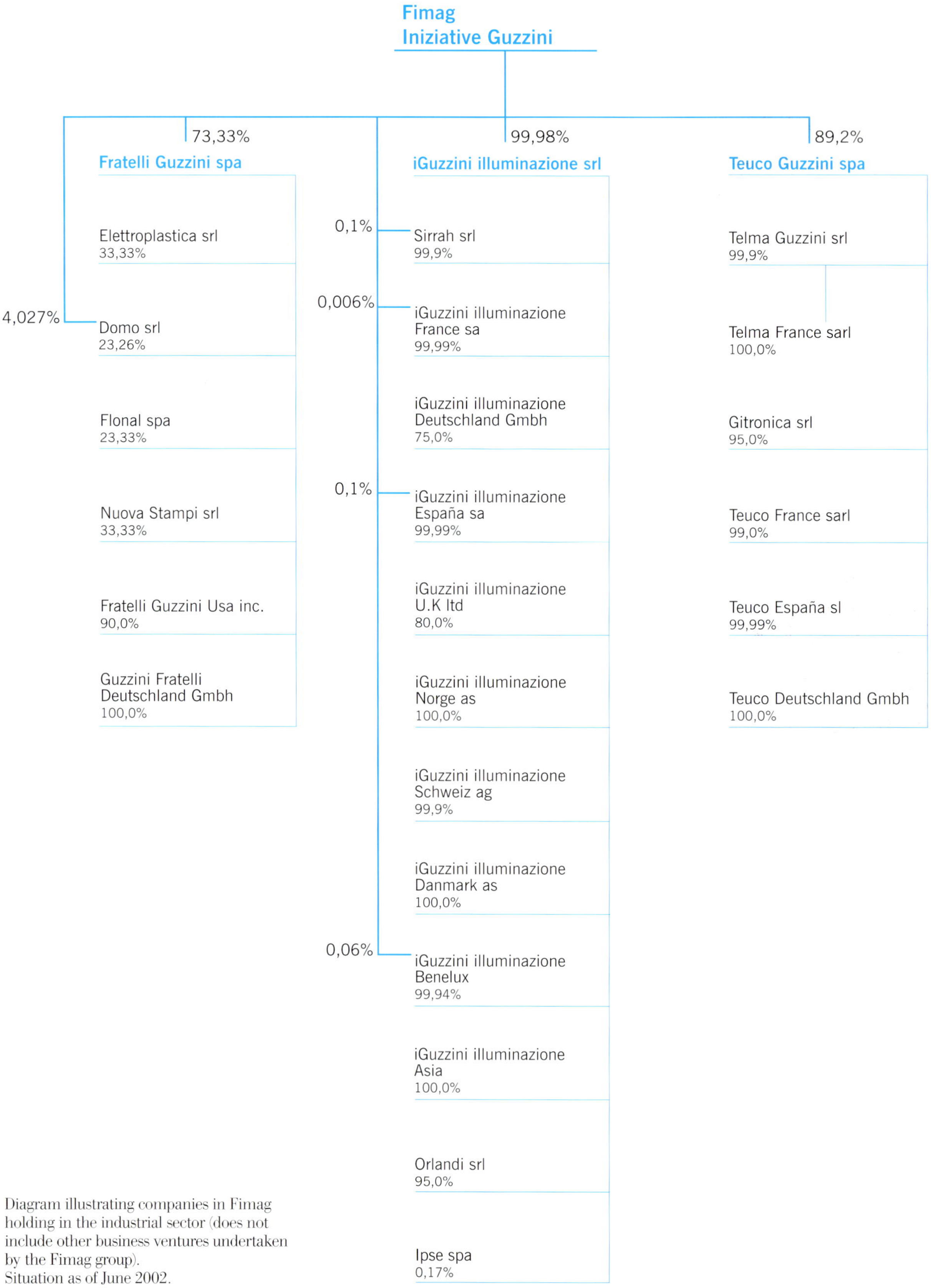

Diagram illustrating companies in Fimag holding in the industrial sector (does not include other business ventures undertaken by the Fimag group).
Situation as of June 2002.

The Archive-museum of Guzzini companies

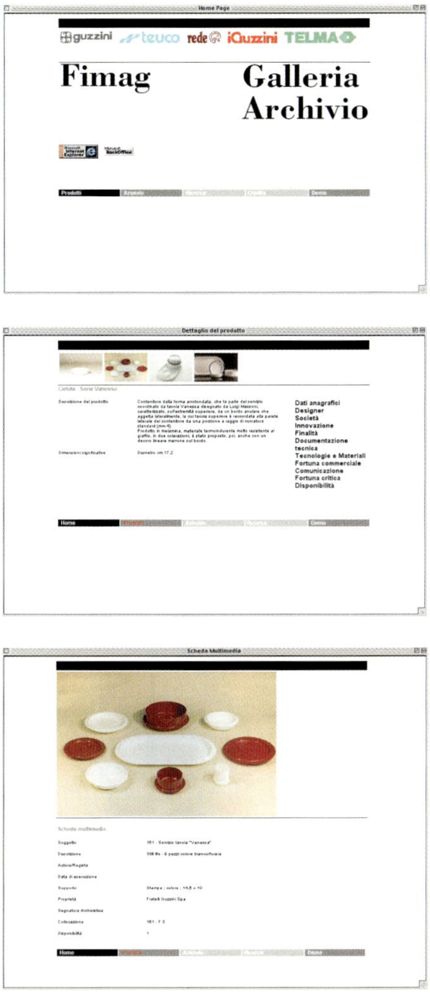

The home page of www.fimag.it and a sheet from the data archive.

In 1995 a brand-new initiative was launched: the Archive-museum of Guzzini companies, a systematically compiled collection of documentation covering all group production history, the results of which are made available to the companies and the public alike.
In the general landscape of Italian industries, many of which allocate resources and strategies to the long-term creation of corporate history museums, the initiative by Fimag , the group holding, has very special characteristics.
The project, designed and implemented by Anty Pansera and the Centro Studi e Ricerca iGuzzini, has a twin objective: to set up a collection centre that helps define company identities, and that of providing researchers, professional designers and the general public with a database of organised and verified historic information.
The definition of a "museum" obviously refers back to the value of historical memory, with all its implications for the present day: to become acquainted with what one was in the past to discern more clearly the paths to be travelled down in the future. "First of all", explains Anty Pansera, "a company archive is designed for archive experts, an in-house service for the company itself. But it also serves a vaster public interested in conservation and above all in the use, as it were, of the past: researchers, designers, Italian but also foreign journalists, and especially students, who are given special attention. In fact interesting cases of synergy have been developed with a number of university faculties, both in the Region and outside it. The company archive is therefore to be seen as a new instrument of knowledge and communication".
One of the initial aims of the initiative was indeed to consolidate an identity of group companies following the complex sequence of events over the years since the beginning of the 20th century, but also to underline the territorial context in which the various group companies have been founded and grown, clarifying their significance and illustrating their characteristics by documents (from sketches, technical drawings, prototypes, through to the end products actually placed on the market, advertising material and documents published in newspapers and books).
A legacy of knowledge that has seen its positive application firstly inside the group, thus fostering the sense of belonging of workers, while making up a permanent training tool of great cultural significance. Secondly, the operation has had an important impact on the exterior as well. The Archive-museum is nothing less than a point of strength for the group corporate image, bearing witness to its cultural background, its strategic choices and the scope of its objectives.
To achieve these ends, the initiative started out with a complex census of existing documents (objects, drawings, photographs, texts), ordered and integrated by interviews with entrepreneurs, designers, technical staff and company workers.
The material, which is still being collected, will make up a permanent corpus, stored in Recanati and already partly on display at the Galleria Guzzini in the town Palazzo comunale. A central aim has been to collect and conserve at least one example of each object produced, of each prototype and preparatory model, of graphic spread sheets, paper and photographic material, company literature, right through to the multimedia material produced during the history of all Guzzini companies.
The data collected are then placed in a database, also located in Recanati. Besides constituting a vital nucleus of documentation available for consultation by companies, the database also provides organised material that can be openly accessed by the public over the Internet, where the Archive-museum is present at the URL www.fimag.it/archivio.asp.
The first stage of works was presented in Milan and Recanati in 1998 during the exhibition "Il bello e l'utile", which offered an initial selection typical of the material available, and in particular of the criteria and results of compilation, consolidated up until 1991 (the year when the group was awarded the ADI Compasso d'Oro prize for its collective activities).

Guzzini designers

Alphabetic list of designers and design studios that have collaborated with Guzzini group companies:

Luciano Altesini
Marco Albini
Emilio Ambasz
Angeletti e Ruzza Design
Gae Aulenti
Shin and Tomoko Azumi
Camillo Baldeschi
Graziella Bemporad
Alberto Berti
Carlo Bimbi
Rodolfo Bonetto
Sandro Brazzoli
Franco Bresciani
Ugo Caggiano
Enzo Calabrese
Giancarlo Capici
Cesare Casati
Mimmo Castellano
Francesco Castiglione Morelli
Piero Castiglioni
Sergio Cavicchia
Pierluigi Cerri
Sebastian Conran
Giuseppe Cormio
Piero Cruciani
Mario Cucinella
Giuseppe De Götzen
De Pas-D'Urbino-Lomazzi
Anna Deplano
Fabiano e Panzini
Forti di Loreto
Frog Design
Norman Foster
Bruno Gecchelin
Nilo Gioacchini
Giugiaro Design
Yoshiko Hasebe
Makio Hasuike
Katja Horst
Martin Hunt
Queensberry Hunt
Charles F. Joosten
Perry King
Cesare Lacca
Ermanno Lampa
Gianluigi Landoni
Latis Architetti Studio
Fabio Lenci
Robin Levien
Ross Lovegrove
Ennio Lucini
Laura Mandelli
Alberto Marangoni
Giorgio Marianelli
Luigi Massoni
Furio Minuti
Santiago Miranda
Pierluigi Molinari
Bruno Munari
Emilio Nanni
Ornella Noorda
Ennio Pasini
Renzo Piano
Francesco and Aldo Piccaluga
Gio Ponti
Carlo Emanuele Ponzio
Ambrogio Pozzi
Thema Progetti
Felice Ragazzo
Karim Rashid
Ambrogio Rossari
Karl Roters
Marc Sadler
Roberto Semprini
Sottsass Associati
George Sowden
Syn
Giovanna Talocci
Dario Tanfoglio
Jun Terao
Tomoko Tsuboi
Ugolini Design
Masanori Umeda
Carlo Urbinati
Jean-Michel Wilmotte

In the words of the protagonists

Documents from the Real/Virtual Archive of Guzzini companies
Interviews by Anty Pansera

The Guzzini family

Raimondo Guzzini in the words of his brothers

Giovanni:
"Raimondo is the brother closest to me in age, because he's the second-born, while I'm the first. We lived close to a Passionist monastery, and Raimondo felt the need to become one – a Passionist – himself, so his monastic life began. After the war period however, he no longer felt that he wanted to be a Passionist, so he resumed his studies in Macerata, enrolling in an accounting institute, which he completed when he was 19-20. He was an extroverted lad, likeable and open, with a definite wanderlust; he therefore decided to spend time with our representative in Naples. After having learned a bit about the profession of the rep, he entered the company and took charge of sales activities. As he toured, like I did myself, he saw products, picking up ideas that he then passed on to me, given that I was in charge of production."

Giuseppe:
"Raimondo, who had taken on responsibilities in the sales division, tried to bring about a few organisational innovations, from the viewpoint of the tools used to penetrate the market. In the meantime he introduced a logo for products, an oval label with gold background and the name 'Fratelli Guzzini' printed together with a crown, to distinguish our products on the market and also to launch a brand policy.
We placed our first advertisements in women's magazines, mainly with the two-colour lines, then we produced a colour catalogue that nicely illustrated our products not only to retailers, but also to end consumers.
Another interesting thing for that time (1960) was the compilation – prepared together – of a price list for retail sales: an important tool, because it stabilised prices on the market. Until that time, in fact, retailers who bought at net prices added the mark-up that they wished, so often there were price considerable differences in prices from one place to another, a situation which damaged trade and even confused consumers. However, by specifying a price to the public, the market was stabilised and the problem was virtually eliminated.
In 1968 we and Raimondo got the idea of establishing a new production business, complementary to Fratelli Guzzini and initially based on imports from abroad – certain exclusive products that could be distributed throughout Italy and to our customers. REDE was thus born.
On that occasion we saw the possibility of bringing together all cousins from Pierino's family (Piero had ten children), my father's family (he had seven), and that of Uncle Silvio, who had three children, even though they were still kids. The project started and progressed; the company started to make its presence felt on the market. The company had a small structure, inadequate in fact for the role that we wanted it to play. In any case it had been launched, and we continued to operate, keeping the books more or less balanced.
Then came 1978, the year when Raimondo, then president of the Cassa di Risparmio della Provincia di Macerata, had a car accident during a trip to Rome and passed away. It was a terrible loss, because many of the businesses operating still largely depended on him, either directly or indirectly. If the truth be known, we had created facilities in the company adequate to ensure management of the various departments and activities.
During that difficult time we brothers just had to grit our teeth and work together even better. Giovanni and I in particular made every effort to keep the venture going, and we did.
Even REDE had to be closed for a spell, because Raimondo was administrator, unlimited partner: the company itself was a limited partnership, and he had all special powers of management. So I personally assumed responsibility for direct management of REDE, together with competent assistants from our company. We transformed REDE into a limited-liability company. However in the light of my deep commitment to Fratelli Guzzini, where a couple of years later I took charge of sales management pending restructuring, I passed management of REDE to another cousin of mine, Stefano. Given that he was still a youngster, I thought that he could gain a bit of training through the overall experience of looking after a small company, and I have no doubt he put it to good use.
In the meantime at Fratelli Guzzini we had obviously continued our activities, together with the contribution of external consultants and design-

Above: Raimondo Guzzini at the Centro Forme (Milan) with Rodolfo Bonetto (September 1969).
Below: the Guzzini brothers. From left, Giuseppe, Virgilio, Adolfo, Giovanni, Raimondo and Giannunzio (October 1967).

Giovanni and Raimondo Guzzini at the Harvey Guzzini stand, 4th Furniture Fair (Milan), September 1964.

ers, who helped transform the internal corporate structure to raise both its efficiency and professional standards.
Even though we had already discussed the feasibility of a family holding, the issue became particularly relevant after Raimondo's death, which in one way helped to accelerate the process.
Fimag is the financial holding of our family. The name is composed of the initials of the words Figli Mariano Guzzini : Fimag, in memory of our father. It was created with the objective of administering company shareholdings that we each personally held in the various Group companies.
The project itself had a rather long gestation period, because we had to persuade everybody, including Raimondo's family and heirs, that entering a holding would certainly not worsen their situation, generate risks of economic loss or less power, or preclude the possibility of entering the company. We thus succeeded in setting up personal and family shareholdings in units. I was very active in the establishment of the family holding, so that it would operate in the interests of family unity and assure efficient running of group enterprises."

Virgilio:
"Thanks to his sensitivity, Raimondo felt the benefits of acquiring new experiences. It was autumn 1957, and Raimondo started up a small enterprise to produce enamelled copper furnishing items. He asked me to become a partner in the venture, because I have always had a natural instinct for tools, materials, processing methods – all things that Raimondo lacked."

Adolfo:
"It was this brother of mine – unfortunately destined to pass away prematurely – who made our business known beyond the boundaries of Recanati and the Marche Region. The industrial project that he set himself was to ensure that the family-based dimensions of the company developed in line with business logic.
The industrial model that my brothers and I favoured was based on confidence in the growth and development potential of the companies, thanks to the industriousness of the family, but also to the resources that the area offered.
We needed to penetrate new markets, new distribution channels, produce objects for a medium-high market. And to do all this, Raimondo sensed the need to capitalise on the technological know-how developed in particular by Giovanni and Virgilio, with the aid of some designers. Actually the designers were the ones who helped build our image, helped us grow.
We brothers knew how to take advantage of the remarkable openings afforded by Raimondo and the enthusiasm typical of certain illuminated entrepreneurs of the sixties, exemplary in the Marchegian landscape, but we were still in a preindustrial stage.
Raimondo showed us the importance of design, of communications and constant attention to market evolution: he planted the seed.
Today, when I read over the history of our family and our companies, I can appreciate how, with my brothers, each of us with his own role and specific human and professional talents, we managed to make a real contribution to the fulfilment of the industrial project imagined by Raimondo, but that he was unfortunately not destined to see."

Giovanni Guzzini

One day in 1941, Mariano Guzzini said to his oldest son Giovanni: "You have to come over and give me a hand, because all the sanders have gone". At that time all products were hand-made, and Mariano knew that Giovanni was good with his hands and had an industrious bent. Giovanni, besides working beside his father in production, given his inclination for design, also enjoyed sketching his ideas for new products. "I was good at drawing when at school, so whenever I had a spare moment – especially when we were planning something new – I would first do a few sketches on paper, then once satisfied I created models.
Once the sketches were done, I would prepare exact templates to create the moulds; when it was possible, I would do it directly. In those times there weren't any special machines to make copies, instead I worked a hardened block of rubber to the required shape and thickness. I designed for example some cruets, which we crafted with raffia handles, or small trays for oil and vinegar bottles. We didn't have designers for our

Above: opening of the new Fratelli Guzzini site in C. da Mattonata, Recanati, 1970.
Below: Giovanni Guzzini accompanies a Chinese delegation during a visit to the company production departments, April 1986.

products then; we did everything in-house, with the ideas that Raimondo or I had. We would make three or four articles of various sizes, then market them.

Investigating and selecting material types was a vital step, because it also stimulated our sensitivity to the material itself and the ways it could be perfected." Giovanni today recalls that the Guzzinis were the first anywhere to work acrylic into sheets, starting out with transparent objects, then coloured ones. The first injection machine was purchased in 1950.

At a certain stage, the Guzzinis crossed tracks with Gaetano Clementi, who brought to Fratelli Guzzini the technology for making twin acrylic sheeting. The results achieved led to the development of coloured products (red, yellow, orange, green, blue), then an exclusive to Fratelli Guzzini. The company thus gained its reputation for originality, unmatched by the competition; it was therefore ready to take on all markets.

"Besides sheets, some manufacturers of semi-finished articles eventually came up with acrylic tubes. We grasped the opportunity to commence working with this new form. We bought tubes of different colours, then decided on which to use; we particularly favoured working with red, yellow, smoke grey and green. Thanks to the tubes we began to broaden our production of carafes, that couldn't be made with sheeting, because they were too deep and the sheets would over-extend. We also used the tubes to produce ice buckets and glasses.

The development of Guzzini initiatives was often a consequence of relations with suppliers, such as German company Rohm and ICI. Our relations led to the exchange of information and expertise, to the discovery of innovative materials, ideas, shapes, and of course the ability to elaborate on them. We were always able to invent new things, but always in harmony with current tastes.

We had to progress from the time when we youngsters decided on the products then made the moulds, as we continued even then. We needed to go beyond activities totally designed and carried through in-house. So we decided to contact a few architects, some through the friends we had.

The first was an architect from Rome, Cesare Lacca, then the second contacted to work with us was Charles Joosten. Together we designed an attractive set of whisky glasses, featuring 5 cm thick bases, hollowed then painted black internally.

Thanks to Adreani of ICI, the company made the acquaintance of a young designer named Luigi Massoni. The contact was made by Giovanni and Raimondo in Milan, then they invited him to Recanati. Massoni convinced the company of the importance of using the skills of trained, modern professionals, particularly important in the field of design and advertising.

We together produced the famous Bolo quadro, which came in different sizes, from 12 cm to 26 cm at the side. It was an article for the home, an attractive container, different from previous models (which were always round, oval, etc.) given that it was based on a squarish shape. It was one of the most successful articles for the Homeform brand.

Besides these innovative containers, there were also trays, salad sets, centrepieces. The US market showed a lot of interest, with a high demand for large containers and trays, giving us plenty of work.

Homeform came at a time when two-colour articles were on the way out. The new material, however, was behind the continuous surge in production in various other countries. Homeform brought an innovation in material, which until then had always been polished; but our articles had a satin finish, inside and out.

Other designers entered after Massoni. To see the fruits of your labour – intuition, study, research, laboratory testing, material analysis – being copied identically might be annoying, but at the same time it is indicative of one's own quality, success, a vindication of operative choices.

All this reminds me of our partnership with Bonetto, whose first work in the late sixties was a bed tray. We subsequently commissioned him to design a pastry tray, which has since become a market classic. Despite being protected by a patent, it has been copied all over the world, including in Germany and Italy.

There was a time when Giovanni and Raimondo considered adding a sector to Fratelli Guzzini for the production of lighting accessories. But the energy and facilities were needed for the household articles area, which was enjoying quite some suc-

Marzia Sabbatini Guzzini and Giovanni Guzzini at the opening of the Galleria Civica Guzzini, Recanati, during the exhibition *Il Bello e l'Utile*, June 1998.

cess. So in 1958, Raimondo and Giovanni (Raimondo and Virgilio had already set up a separate business for the production of enamelled copper items) decided to establish a venture for lighting articles. Giuseppe and Giannunzio immediately entered the company, while Adolfo only joined on the completion of his studies. Giovanni designed the first collection of lamps manufactured in the workshop that he had installed in his own house. Three years later the first production site was opened, the precursor of what would later become iGuzzini illuminazione."

The suggestion that Giovanni gave to the new Guzzini generations was to be receptive to the signs of the times, then to compare them with their own ideas to see whether the latter were well-founded or actually in conflict with others. He also urged that the creative force inside the company be maintained by combining it with external sources, with input from consultants and outsourced collaborators.

What was, however, vital was that no opinion, even from a renowned technical expert, should be accepted passively; it should rather be compared to one's own ideas, to the objective requirements of the company and the market. All the knowledge possessed – whether technical, aesthetic, personal – gives rise to a synopsis that is reflected in the final decision, which is the responsibility of the person at the head of the entire company".

Virgilio Guzzini.

Virgilio Guzzini

"First of all I have to thank my parents for teaching us all the value of honesty; they also reiterated that our strength would be even greater if we stayed together. And years down the track, their advice remains valid, because our strength is in our unity.

1955 was a fortunate year for me, because Clementi taught me and my cousin Alberto Guzzini (uncle Pierino's son) how to depolymerize and distil the sheet cuttings used during the processing of our articles, until we obtained an acrylic monomer, PMMA, that was sufficiently pure for the production of more sheets. At that time acrylic was as expensive as silver, so recycling of waste material generated huge savings.

I was subsequently put in charge of production of colour acrylic sheets by superimposing casts, used to create food containers, white inside and coloured outside, an unqualified commercial success.

Going back to the early fifties, the second of the Guzzini brothers Raimondo began to work as a representative of Fratelli Guzzini products. During his extensive business travels throughout Italy, he had the opportunity to observe an abundance of new things in a thousand different fields, to develop a sensitivity to all types of innovation. His finely-honed sensitivity allowed him to perceive the possibility of acquiring new experience; it was autumn 1957.

Raimondo started up a small business to produce enamelled copper artistic objects. He asked me to take part in the venture by becoming a partner, given that I have always had a natural instinct for tools, materials, processing methods – all things that Raimondo lacked. And since he was often away on business for Fratelli Guzzini, Raimondo needed to be able to count on someone in charge of coordinating production, shipping and management in general. This was the beginning of what would become Harvey Guzzini di R. Guzzini & C. s.n.c. – and virtually at the same time, my brothers and I decided on a new lighting product line that combined enamelled copper and acrylic.

In the meantime, Clementi, who had considerable (even foreign) experience in the chemical field and had in 1958 opened his own small company in Pomezia (Rome) to produce acrylic sheeting for luminous signs and advertising billboards, offered to personally enter the partnership. In September 1961 I therefore moved to work in Pomezia.

My experience there gave me the chance to travel extensively in Europe and Asia; furthermore, I used the quality of the material for special activities. I commenced operating with scientific apparatus for CNR in Frascati, particularly on light guides, because acrylic with a thickness of 40 - 50 mm or greater is perfectly transparent – even better than glass, though a lot less fragile. I worked with major Italian and foreign furnishing names: with Frenchman Jean-Claude Farhì we furnished the Gunter Sachs apartments in the Palace Hotel of St. Moritz, and those of his brother Wilhelm in Rome in piazza Santi Apostoli at

Virgilio Guzzini receives the award to Harvey Guzzini and Fratelli Guzzini during the Europremio furniture and furnishing fair 1970 (April 1971, Rome).

Palazzo Odescalchi. I also recall a large number of projects with the architect Anna Laura Angeletti from the Studio Arredamento Navale.
I continued my collaboration with artists, such as Gino Marotta, as they crafted their objects from acrylic. At the end of 1969, the Embassy of France in Rome thus summoned me to work with two artists, the Poirier, to prepare the French pavilion at Expo 70 in Osaka: we created a series of human figures from acrylic, each of gigantic proportions, approx. 300 x 200 cm.
My visit to Expo 70 with my brother Adolfo gave us a lot of food for thought. In a Japanese health club, I had my first experience of a whirlpool tub (invented by the Americans in 1968), which captured my imagination. The following November, during the Interbuilding Fair in Birmingham, we met the owners of a company that moulded bathtubs from acrylic material, P&S. That meeting was a real shot in the arm, that set my imagination racing, a process that culminated in late 1971 – early 1972 when I returned to Recanati. In fact on January 20th 1972, my brothers and I incorporated Teuco Guzzini.
I can say without fear of contradiction that Teuco in the seventies – just like Fratelli Guzzini in the thirties – was a pioneer in the use of acrylic and plastics in general in the industry, indeed becoming a model enterprise, setting new standards for our competitors.
We commissioned P&S to craft the first two moulds; a shower tray and a bathtub. On the tray we constructed the moulds for walls and the colourless transparent doors, with excellent results. In a relatively short period of time we succeeded in presenting the first shower and bathtub prototypes at the Mostra Convegno in March 1972.
It was a great success, despite the scepticism towards plastic as a suitable material for bathroom fittings. Guarneri, who at the time was the largest wholesaler in the Milan area, even asked me if the material was resistant to hot water!
The introduction of the multi-function shower definitively assigned Teuco the role of market innovator and trend setter.
When the decision was made to go ahead with the Teuco project (the name had been coined, just like that of Telma later on, by Carla Caccia), Adolfo began to transfer the expertise of collaborators acquired at iGuzzini from the ADVEMA agency (then responsible for iGuzzini advertising) to a designer whom we considered the most suitable for the project: Fabio Lenci.
The sales side of the activity also had to be developed. Thanks to Adolfo's personal relations with ISTAO, we contacted two young graduates from there: Antonio Santi, today general director of iGuzzini illuminazione, and Antonio Renzi. Renzi chose to work with Teuco and is still with us, in fact I almost consider him a seventh brother. It is no accident that he is also a Teuco partner.
Despite the problems encountered during the first ten years, we were optimistic, and in fact Teuco enjoyed enormous commercial success. In this light, in 1985 we purchased a first lot of industrial land in the Municipality of Montelupone, to build our new production site. We are still there, actually the facility is still growing.
Returning again to the early eighties, I ran into Maurizio Adreani, whom I had known since my younger years, because his father was the dealer for ICI (Imperial Chemical Industries) and supplied acrylic sheets to Fratelli Guzzini. Maurizio proposed that we test a brand-new material, produced by ICI, called Asterite, which we subsequently used to make kitchen sinks, basins and bath tops. It was thus that in autumn 1981 Telma Guzzini S.r.l. was born; today it is a company with its own consolidated presence".

Giuseppe Guzzini

"Our companies – which are family enterprises comprising brothers and sisters, uncles and aunts, cousins – need unity to succeed, to progress smoothly. Unity implies the prevention of conflicts, favouring agreements and achieving a common stance among all the people concerned by the family business, while respecting the principle that a company is to have a structure that is efficient, innovative yet autonomous, distinct from the family, with its own rules that we above all, as owners, were required to obey, to reassure the company, the people who work for it and our families that, at the end of the day, new activities also depend on profits.
In 1954, once my accountancy studies were over, I entered Fratelli Guzzini, obviously in ad-

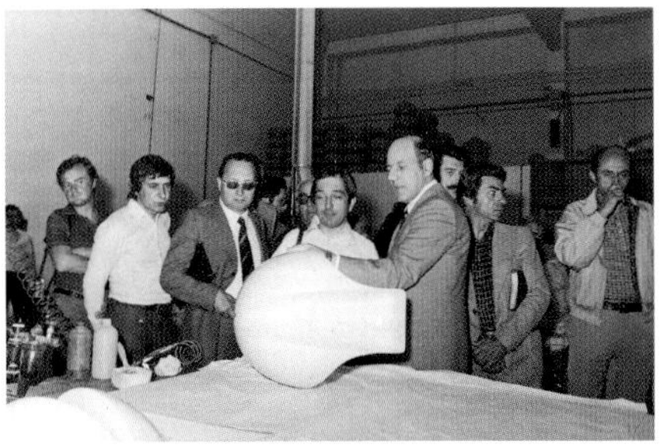

From top to bottom: Giuseppe Guzzini at the Centro Forme (Milan), September 1969.
Giuseppe Guzzini visiting Harvey Guzzini production site with clients, 1974.
The Guzzini brothers at the inauguration of the exhibition "Il Bello e l'Utile" at the Milan Triennale, May 1998. From left: Adolfo, Giuseppe, Cesare Annibaldi, Virgilio, Giovanni, Giannunzio.

ministration, working beside my uncle Pierino who until then had looked after the books by himself. I was assisted by my cousin, Ines.
When I entered the firm, I endeavoured to become acquainted with the situation and the current corporate organisation. Then on the basis of these studies, I illustrated to the partners the need for an organisational, accounting and administrative structure by which the company could have every week (or at the most every month) an orderly, updated, freely-available report on company matters.
Up until then, management was conducted manually, in a simple, rather makeshift manner. I eventually succeeded in persuading my father and uncle Pierino of the need to purchase new Olivetti equipment to render account-keeping practices more efficient, of the utility of accounting tools such as day books and ledger cards to help compile a monthly review of the company situation. These new management tools were subsequently introduced to organise a system for checking product costs and profits. I thereafter tried to transfer the benefits of this experience whenever a new activity was initiated: organisation and administrative management, relations with credit institutions, contract definition, the management of new statutes and agreements among partners, the incorporation of new businesses.
My uncle Pierino had a particularly large family – ten children in all, four sons and six daughters – and he was clearly quite concerned about their futures. In the same way my father was concerned about the future of his seven children, and uncle Silvio about his three. The only strategy possible to create new prospects and openings for the younger generation was to create new activities, either inside Fratelli Guzzini or outside it.
In the face of these concerns, uncle Pierino unfortunately believed that our activity for the production of articles for the home and relative channels of distribution – the best clientele portfolio in Italy and la Rinascente in Milan, an excellent customer – were sufficient to guarantee both the family and the company seamless progress into the future along the same path.
During the fifties, we experienced a positive patch thanks to our two-colour products made with the sheets produced in our chemical plant. Towards the end of the decade however, the market for these products began to wane. We, and in particular Raimondo, increasingly brought in products made with the injection system by our competitors. We needed to change technology and make investments, otherwise the company ran the risk of falling behind, or even collapsing.
We were already acquainted with the new injection technology, having bought an English Windsor injection press in 1953 to mould goblets. Unfortunately it was not immediately perceived as the new strategy capable of guaranteeing more rapid development to the company, an extra speed to help us grow. Uncle Pierino was reluctant because, as I said earlier, he was very confident the market for domestic articles would hold true. He was also afraid that the medium-term loan that we would have to take out, and of course the mortgage on company assets required, could damage our general bank rating. But in actual fact the banks would have appreciated the initiative given that it was a sound one, that would not weigh the company down with immediate debts, that would bring contracts for new investments.
To avoid unwanted fiscal repercussions, this operation first of all forced us to incorporate a regular enterprise; in 1960 we therefore made the transition from de-facto company to an unlimited partnership ("s.n.c"). Secondly, we drew up the articles of incorporation that laid down powers and specified the business's practices.
We thus obtained an initial loan of 21 million lire – enough to build new facilities, buy machinery and make moulds for the first set of products to be placed on the market.
The business developed thanks to our commercial success, so we were fortunately able to draw up new programs and broaden our range of products. The market reacted positively; while two-colour articles were still being sold in the most prestigious stores, injection articles were supplied to an even vaster clientele through more widespread distribution channels.
In 1958, after uncle Pierino had turned down the idea of introducing the production of lamps to Fratelli Guzzini, we, Mariano's children, incorporated the Harvey company for that precise purpose. This decision generated some degree of internal friction that we tried to alleviate by a logic aimed at avoiding disruptions in the family and

Giannunzio Guzzini with the *Greenwich* designed by Norman Foster in 1999.

the company, by developing new activities that would involve other younger family members, like uncle Pierino's children themselves. We thus reached an agreement: acknowledging the existing businesses, i.e. Fratelli Guzzini and Harvey Guzzini, the principle was established that anyone, whether individually or as part of a family group, had the right to set up new activities on the condition that they were not in competition with existing enterprises. In this way uncle Pierino's four sons got together and started up Acrilux for the production of lighting screens and acrylic ceiling bowls, made with advanced sheet and injection processes, destined to become a market leader in the industry.

To conclude, I can state that every time a development has been pushed through, rapidly and in the face of opposition, I have always tried to reach an agreement and to create the conditions so that, rather than allow a rupture in the wake of events not universally accepted, discussions were possible to prepare a negotiated solution for future development, in the interests of the whole family and relative companies. In other words, I've played the role of a go-between to coordinate between new activities as they gradually emerge. The strength of these six brothers has been their unity, their complementary professional roles, the commitment to objectives common to both enterprise and family".

Giannunzio Guzzini

"When I turn back and look at my entry into the workforce, I have to admit that I would never have imagined seeing that 14-year old boy, who grew up in the workshop next to his father, an authoritarian but irreplaceable teacher, come so far.

After receiving my diploma, I was introduced to the industry as a labourer next to my father at Fratelli Guzzini, and as such I was required to observe all the rules, including those covering working hours, which at the time varied between 10 and 11 hours a day.

During the same period, my three elder brothers used their free time outside Fratelli Guzzini to set up their own businesses. The most important in terms of the developments that ensued was based on the production of Plexiglas light fixtures. I believe that one of the best days in my parents' life was when construction works were commenced on the Harvey production site; but besides the sheer satisfaction at construction, I think that the real gratification came at seeing their sons united in that venture.

In the early days of Harvey, Giovanni wanted me as his assistant diesinker, an activity that I conducted during my free time from Fratelli Guzzini. I would work the wood block with a hammer and chisel, to create the form of a mould used to cast a luminaire or a part of it.

After a few years, I started to grow restless, trying to assert my professional skills, until the day I worked up the courage to tell my father about my intention to ask my brothers for a position at Harvey, which I officially entered in 1966.

Alongside Adolfo, who had already been working there for several years, we dedicated our undivided attention to Harvey, while my other brothers continued to make their own contributions, reconciling them with their commitments to Fratelli Guzzini due to their increasing responsibilities there.

Adolfo and I always shared tasks, following our natural inclination consolidated over many years in the company. Adolfo looked after the commercial and organisational development of iGuzzini, the company image, with his natural gift for anticipating market trends. He specialised in selecting designers, in creating the sales and marketing of the business. And since the very beginning, Adolfo followed product developments, constantly intensifying collaboration with architects and designers. I too developed excellent relations with them, in the design of new models, to reconcile aesthetics with technical facets of the product.

I was equally dedicated to the design of certain machines, to help improve processing in line with a more industrial approach.

The seventies were a time of patents, some of which I can claim responsibility for, in particular the entire project for injection-blown globes, designed at the urging of a friend and competitor who purchased this type of product in the US, where it was made by rotational moulding. I had memorised the idea and was looking for the right context to start more in-depth studies, but using a more functional production technology than our own, able to ensure a higher-quality product. One

Above: 1st National Conference for Light Engineering Specialists, organised by iGuzzini, March 1977.
Below: Ceremony on the appointment of Giannunzio Guzzini to the presidency of CELMA (European Federation of National Manufacturers' Associations for Luminaires), February 1995.

day I happened to be handling some cylindrical components of a Fratelli Guzzini cache-pot, made by injection technology. I picked up three or four of these articles, then started some rudimentary blowing tests of the cylindrical volume, which when expanded in free air formed a spherical shape, that obviously split when the press was opened.

Adolfo and I concurred that we should continue the development of the product, by building – even inside the company – machines based on a technology for the production of globes in large quantities, thus at an industrialised level; we subsequently designed two more technological methods that are still being used. This product, which over the past 20 years has generated a remarkable turnover for the enterprise, is without doubt the reason behind the current dimensions of iGuzzini.

But the attention of production was not limited to globes; in the mid-seventies we in fact came to the almost simultaneous decision to drop decorative production to concentrate on lighting engineering. Adolfo was the first to be convinced of this decision, in virtue of the impressions gathered during his frequent business trips that market demands were heading in that direction. I myself had to make a number of trips outside the company to become better acquainted with metal processing technologies that were totally new to me, such as cold forming of sheet metal by drawing, the turning of aluminium, steel and stainless steel sheets, trimming and bending, pressure die-casting of aluminium, magnesium and zamak, shell casting, different types of mechanical processing with conventional then NC machines.

Over the years the company has grown, while many of the activities that I myself carried out are now performed by a group of highly-skilled and professional technicians, whose dedication to the enterprise has allowed them to attain the highest levels of responsibility.

After years of working side-by-side, I could easily give a detailed description of each group member, but I will only mention a few (apologising to the others not mentioned due exclusively to lack of space): Antonio Santi, Mario Sampaolesi, Franco Nibaldi, Pietro Clementi, Sandro Rita and Piergiovanni Ceregioli. This open approach, from which no-one showing their commitment to our companies has ever been excluded, is certainly one of the main strengths of the enterprises run by our family, that of Mariano Guzzini.

As I mentioned earlier, my work has allowed me to develop excellent relations with both our suppliers and our competitors, giving me an in-depth knowledge of this particular industry. Indeed these relations are behind the formation in 1969 of a group of friends/competitors, the Italian Association of Luminaire Manufacturers (AIFAI), today Assoluce, part of Federlegno. Both Adolfo and I have served 8-year terms as AIFAI chairman.

This commitment to the development of the industry has continued even more intensely in ASSIL, an association that is part of ANIE, of which I am currently chairman".

Adolfo Guzzini

"In 1964, a few years after my graduation, my brothers appointed me to a full-time position at Harvey Creazioni, in charge of organisational, administrative and commercial matters. Together with Giannunzio, who arrived some years later to look after the development of the technical facets of production, we brought the enterprise to its present-day dimensions.

The mid-seventies saw the emergence of a more discerning public for contemporary products, a market that could not be entirely satisfied by Harvey. In 1967 I therefore decided to launch a new product line under the brand name DH-Design House. I called in Luigi Massoni to act as art director; he had the task of coordinating the range, as well as inviting other designers: Ponti, Casati, Ponzio and Lenci, to name just a few.

A vital contribution was made by Massoni, so much so that at the end of the sixties I proposed Massoni's services to many of my friends in the Region, because he had succeeded in imbuing the concept of design not only in products but in sales organisation, corporate image, as well as communications, from catalogues to advertising. The Centro Forme that we opened at the time in the heart of Milan, in via Durini, was a venue where the company organised events and presented products, years before the idea of the showroom was actually fully developed – an aspect of our enterprise culture that our companies

Meeting with Renzo Piano for the definition of the *Lingotto*.

believed, then as now, of strategic importance to overall development.
I realised from the very outset that the company needed to see the on-going development of the market as an opportunity for growth, rather than something that disrupted a situation of balance. A leading firm aspiring to maintain close contact with the market had – and has – to make investments at least equal to or even greater than those necessary to ensure technological upgrading and adequate customer services. In this case the entrepreneur must have a broad outlook on external factors, be confident of the strength and determination of themselves and their workers.
During that period I identified a further aspect of strategic importance to growth, something subsequently adopted by other group companies: exports to the more prestigious international markets, the key to quantitative and qualitative growth of the enterprise. Unlike other firms in the industry that counted on a single agent in charge of the same market, we advanced the idea of a new system of distribution based on exclusive national dealers who handled products that were complementary to our own. So as early as the late sixties we had a network of highly-integrated trade organisations in all major European countries; in this way we could acquire in-depth knowledge of each market thanks to our direct presence, besides laying the foundation for our present-day branches. As part of this European strategy, in the same period we also started up a joint-venture with Meblo for the production and distribution of our products in East European markets.
Our acquaintance with foreign markets proved useful in 1974, when the repercussions of the oil shock hit our industry and suddenly overturned our plans for development. We faced a choice: either scale down our industrial ambitions or diversify production. I suggested the possibility of diversification into architectural lighting systems, something that I had learned about on other markets but still did not exist in Italy. Architectural lighting is designed not so much for domestic applications as for public areas where people gather: shopping centres, hotels, offices, museums and so on. Lighting fixtures are normally selected by the person in charge of the overall architectural project. Thus an approach to products and the market completely different from that of home lighting, and something that required a veritable transformation of the entire company.
My brothers accepted my proposal, despite the awareness that diversification involved enormous investments and corporate refinancing, employing all our company and family resources. By the mid-seventies Harvey had become iGuzzini illuminazione. We launched our first range of technical lighting equipment; though I had a clear idea about our products, I realised that we urgently needed professional designers and greater production facilities. Steps in these directions were rapidly made, thanks also to the enormous technical skills and attention to product quality shown by Giannunzio, whose studies had unveiled new opportunities and both product and processing innovations, as seen in the case of the polycarbonate globes; Bruno Gecchelin's designs were the perfect expression of these new principles of product.
Sweeping changes were made to both technical and commercial facets of organisation, with the collaboration of some of the finest architectural lighting specialists around, including Louis Claire from France, Van Malotki from Germany and Luigi Manzoni, who all trained our technicians. Our sales department for example boasted a new figure – that of sales-technician, whose task it is to respond to the lighting demands posed by architects. At the same time, we invested in communication, which we have always considered strategically vital, in the awareness that the issue of lighting quality was linked to other latent questions; despite their important social implications, these questions had never really been analysed. By the advertising campaigns launched in 1982 on the relationship between light and people, light and architecture, light and the environment, culminating in 1993 with the slogan 'With iGuzzini against light pollution', we were the first in Italy and indeed the whole of Europe to draw attention to the importance of technical lighting, until then the domain of a handful of specialists. We gave support to other companies specialised in architectural design, such as B&B, Anonima Castelli, Unifor, the rebirth of the magazine 'Rassegna' edited by Vittorio Gregotti, not to mention the graphic design project by Pierluigi Cerri, aimed at making a tangible contribution to the

Above: Bruno Gecchelin and Adolfo Guzzini receiving the Compasso d'Oro awarded to the *Shuttle* system, 1989.
Below: Business meeting with Mario Botta, 2000.

debate over the lighting issue. After all, if production is aimed at an expert public, common to all individual national cultures, companies have to invest to ensure that their message is successfully conveyed to the market.

Over the decade leading up to the mid-eighties, the company had managed to become a dynamic presence on the international market, known for the quality of its services and products able to match it with Europe's finest. During that era it had also initiated collaboration with the greatest names in contemporary architecture, including Renzo Piano, Gae Aulenti, Piero Castiglioni, Norman Foster, Jean-Michel Wilmotte, who would design some of our most important luminaires of the nineties.

When I tackle a problem of a single enterprise, I always try to place it in the general context of the group. That's why, in the mid-eighties, besides my commitment to iGuzzini, I proposed that our various enterprises be coordinated by a body referring back to the family, for the elaboration of development strategies. The project, which was initially put on hold because deemed too expensive, was resumed at my instigation after Raimondo's death; Fimag, the family holding, was thus born.

I'm the youngest in my family, separated by 14 years from Giovanni, the oldest, so I've experienced a differing way of thinking, of conduct from that of my brothers. That's how I realised that our family not only had to look after business matters, but also the relations of our generation and all the younger ones, our children and grand-children, by investing money in training and time to help the younger family members become aware of the need for total dedication.

I therefore took responsibility for the creation of the "Patto di Famiglia" (Family Pact), thus defined in 1980, without doubt one of the first schemes of this type to handle such a delicate issue, a recurrent one in Italian industries based on a family business.

The training scheme is designed to prepare members of the younger generation to take up our cultural, ethical and industrial legacy – because the company is not only a family asset, but a social one as well, with a direct impact on the cultural and economic life of this country.

Some members of the fourth generation have already taken on this strategic role, such as my nephew Paolo, who since the eighties has had the task of guiding the sales department of iGuzzini illuminazione, developing its international scope.

When I look at our enterprises, I think that the feature that best exemplifies their history and assures its future is the unrelenting determination to invest in innovation, in the training of our human resources, in the extension of our presence throughout the international marketplace.

Interviews held in 1998.

In the words of the witnesses

Documents taken from the Real/Virtual Archive of Guzzini companies
Interviews by Anty Pansera

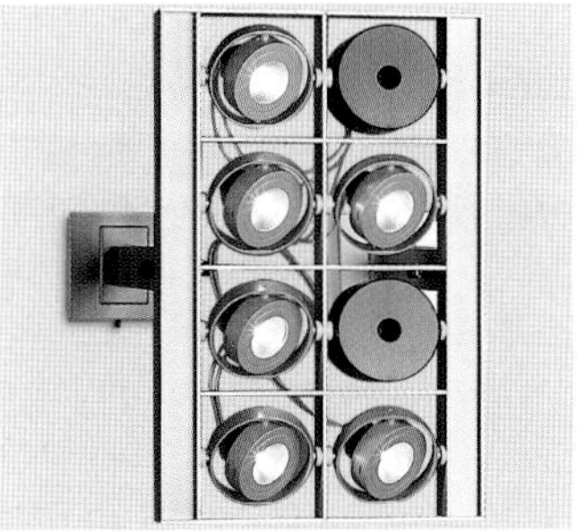

Cestello designed by Gae Aulenti and Piero Castiglioni in 1993.

Radius outdoor luminaire, designed by Piero Castiglioni in 1997.

Gae Aulenti

"The history of the Cestello dates back to ten years ago, the first idea for illuminating Palazzo Grassi in Venice.

The Cestello was derived from an analysis of lighting systems for large public venues, such as stadiums, with reference to the majestic, monumental 'basket-type' projectors, though of a smaller size, adapted to the dimensions, the spaces of Palazzo Grassi. It was a component, a way to keep a set of lamps together, designed to serve the needs of a temporary exhibition.

What was the Guzzini intuition? When the Guzzinis visited Palazzo Grassi, they realised that the Cestello could be used for other applications. Efforts were therefore undertaken to transform this module, crafted in a small number for a specific venue, into a product with a broader spectrum of uses. The Guzzini company used all its expertise and productive capabilities to eventually create not only the Cestello module that is most commonly seen today, but also standing models, that we adopted for example in a room in the Quirinale palace, as well as suspended central modules that we installed in some stores here in Milan or in the restructuring works at Palazzo di San Marino. As problems emerged, we found that the Cestello was sufficiently flexible to adapt to different installation requirements.

The industrialisation of the product proved to be quite an ordeal. Why?

Because it is one thing to handcraft an object, in which for example certain components are made to a minimum thickness, another to manufacture the die plates to make the parts that then have to be assembled. Because the main thing, at least in my way of thinking, is that the object be as inconspicuous as possible.

So Piero Castiglioni and I forced the situation slightly, asking Guzzini to make the modules in two sizes: one for spacious areas, therefore with larger lamps, the other for smaller-scale applications, for example to illuminate paintings in a private residence – so with much smaller components. We succeeded in making both types. The production department naturally illustrates in the catalogue the most common functions performed by the module, while we well know that the Cestello can be used for more specific applications. The industrialisation process in any case adequately answers the demand for ad hoc equipment. This is very important in a process that must take the demands of a broad market, of wide-reaching distribution needs into account, while embracing the potential to produce very special articles.

Such "mobility", as it were, in an industrial firm like Guzzini is something I staunchly defend, even in view of what takes place in other countries".

June 1998.

Piero Castiglioni

"I first met Adolfo Guzzini in the early eighties. Adolfo, who is currently company president, came to my studio several times because he was interested in my work, and naturally to ask me to design his products. At the time – and this situation lasted until the beginning of the nineties – I was working for Fontana Arte, so I had to turn him down, but clearly with no hard feelings. Subsequently we began to design outdoor luminaires, then we started work on the Cestello for indoor applications; he subsequently began his obstinate bid to buy the rights to the product, which he wanted intensely.

Giannunzio was the second Guzzini that I met after Adolfo. I have always appreciated his meticulous approach when designing products. Even today, and though my ties with the company are officially much looser, Giannunzio is the key figure in product design and implementation.

The client is indispensable to a project, as is the person who carries it out, the workers in the case of construction works and industry.

What I am saying is that when we succeed in taking responsibility for the success or failure of a product, we are only actually 50% responsible – half the success or half the failure. The client is essential: think of Florence and what it might have been without the Medici. We believe that the client must be part designer, while the designer must be part businessperson. It is precisely in this exchange of roles that a dialogue that is usually profitable can be played out. In the case of the Guzzini company, I must admit that Adolfo is a bit of a designer, as is Giannunzio, while I, due to my years of experience, have been a bit

Trimmer system, designed by Gae Aulenti and Piero Castiglioni in 2000.

Mylife bathtub by Frog Design.

of a businessman; so we understand each other's problems well. For example, when we were making the Cestello, my concern was that of having to tell Adolfo, "Look, this product might be successful regardless of its applications, of its performance, if it is cheaper or the same price as the track systems; because if that's not the case, we'll leave the market, and it's pointless to make a product that doesn't have one". The problem of economic costs is therefore one that concerns the designer as much as the businessman."

May 1999.

Frog Design

"For some time now we've been studying the evolution in the concept of comfort: not only more refined functions, but ones that are also easier and more convenient, increasingly-natural shapes and materials. We would like to create elements for a setting that helps the person recover their resources, even perhaps by evoking lost or forgotten sensorial experiences.
In 2000 we designed a complete range of bathroom products, including a spacious bathtub. We were on the lookout for someone able to make the concept a reality. We were looking for a company with production and technological know-how that could charge the project with quality details and cutting-edge performance, who shared our dedication to the pursuit of comfort.
We soon discovered that Teuco was the only industrial venture around that was able to share this experience with us".

May 2001.

Ross Lovegrove

"I think that I am constantly searching for an intimate relationship with the things that surround me. All the things that we live with are in one way tools, and often I realise that, for me, beauty lies entirely in function: beauty is in the gesture, not necessarily in the way you look. It is a sensual dimension too, the way in which you let yourself be touched. Within the context of the work that we are doing with Fratelli Guzzini, we have tried to use plastic to elevate the condition of the object, taking it beyond the banality of something anonymous, one of the many in life.
Plastic's beauty lies in its being the mother of all materials, in the sense that it can become anything you want: it can be wood, it can look just like cement. But as far as it concerns me, I look for honesty in the material, just like some designers look for weight or depth. I strive to make a minimum use of it, or in any case so that it has its own identity. I love the sincerity of materials and it is my firm intention to let them live, let them breath in a way that I define as 'new nature'.
I have a fantastic idea in mind, and one of the reasons why I seriously intend pursuing my collaboration with Fratelli Guzzini is that I am one of those people who, when they get out of bed in the morning, look at the future. I never think about the past, and I don't set myself limits.
Today we have the opportunity to do so many things. My dream is to create what I call "Plastic Villa". I want to build the first house made entirely from plastic, and I want to build it here in Recanati. Whenever I see Giovanni Guzzini looking out his office window and saying: 'See all this land? This is my land!' – a bit like JR in Dallas – I keep on thinking: 'Why don't we build a new Vitra here in Recanati? Let's build surprising buildings!'.
I went to visit Teuco and I think of those remarkable bathtubs, the integration of shapes: the walls that become a tub, then a seat, then they become... life. That's it, for me all that is absolutely natural. I like the idea that the machine makes the product, then the pieces are put together and the result is lovely, clean, light and modern.
I just can't think of anything else! Maybe I should get Domenico Guzzini drunk, try to get his signature on a cheque and start work on an extraordinary plastic villa. I really think I'll do it.
The reason I love Europe is its wealth and diversity of culture. I like the idea of the pursuit of the physical, a language suited to different markets, because cultural differences exist and it is precisely this that makes the world so beautiful. I think that Fratelli Guzzini is well-positioned from this point of view, also because of the type of people who work with them, crafting things that have an intrinsic utility and that are instruments to be held. What I mean is that my hand is like yours and like those of men in the Far East, except perhaps for the size. Hands are always the same and

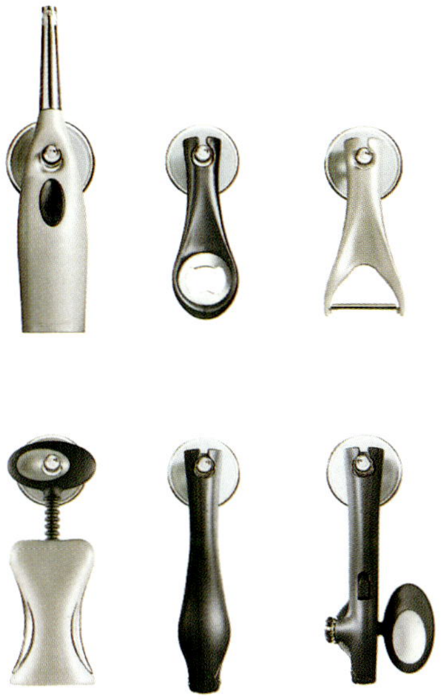

Handy's accessories designed by Ross Lovegrove in 1998.

do the same things. The evolution of man and his body has taken a long time. So if adequate, suitable objects that give the impression of being ergonomically beautiful are created, they can only be improved by using new materials or new production technologies. That means that in theory we can design a product today that will last ten or fifteen years, because it works well and is a fine object. My intention is precisely this: to make things able to go around the world, globally but without arrogance.
I'm very satisfied with the kitchen utensil project for Fratelli Guzzini, because it matches up well with some of my research studies in other areas. When I design these objects, I always think of them in a certain context, not as isolated things, also because I have the opportunity to work all over the world, with different companies, and this gives me an overall insight into what happens in that setting. Working for Fratelli Guzzini has been a golden opportunity, because they respect my ethos and they encourage me to design new things. So we're still actually at the beginning: we're doing other things that will in any case have their own use, consequently the people will buy them because they need them.
When I was asked to work on these objects for the kitchen, I took a stroll through the company, wanting to personally touch the history of Fratelli Guzzini, the first utensils made from horn: man who tries to be a machine, but is really an artisan. I also like the economical character of the material, because when you simply get a surface with a certain thickness from a material, it really is the minimum necessary. So I thought: "I'd like to make a connection between the origins of Fratelli Guzzini and the possibilities of today". That's why my articles resemble bones: as bones are their origin".

June 1998.

Renzo Piano

"My adventure with iGuzzini began quite by accident. Around 1985, we were in Turin, in the worksite of the Lingotto, we weren't sure how to make light, we didn't like anything, we weren't satisfied. Then one day Franco Nibaldi – he's the culprit! – dropped by and left us with a Guzzini lamp, that initially didn't suit us; then gradually, between a joke and a more serious moment, we started to play a rather unusual game of ping–pong with Adolfo Guzzini, between designer and manufacturer, ideas flying back and forth, discreetly, without fanfare ... we were fine-tuning a concept. After two or three months of 'giving and taking', we finally came up with this lamp, our first produced together.
I am proud of it: once, during a small exhibition at the Lingotto, someone came up to me and said, "this lamp was of course already here at the Lingotto!". I answered, "Yes, it was a piece of the Lingotto !". This is a joke I like to tell, because in actual fact every object holds within it a set of allusions, a story, something imaginary.
Then we made many other things, more or less successful; we still are, but still with this approach of continuous 'give and take'.
In 2000 there was what we called Le Perroquet, a name that came about as a joke. We made it for Beaubourg, because we didn't know how to illuminate the Forum and the exhibition on office work. So we said, "well, what we really need are some parrots suspended in mid-air, objects that drop down". We eventually designed a stand with this object suspended on two wires, with the power line in the centre. At Beaubourg the wires are long, there are suspended objects occupying the space, inhabiting it you might say, joyfully, with a type of taste for allusion, jest, a certain irony.
We'll continue with the Guzzinis, who knows what will be next – a surprise! It's a partnership that I like talking about, to testify, because it's discrete, with great programs or impressive declarations – rather a matter of everyday life, of 'give and take'. Occasionally we have a tiff, but that's normal ... but very, very few, and always settled. That's one thing I'm thankful for, not only to Adolfo Guzzini, the person I meet most often, but to everyone else from the 'big family'".

January 2000.

Ettore Sottsass

"The issue of plastic materials is vast, because plastic today invades every aspect of the living landscape, as it were. Over the last few years it

Le Perroquet Spot lighting system, by Piano Design Workshop (2000).

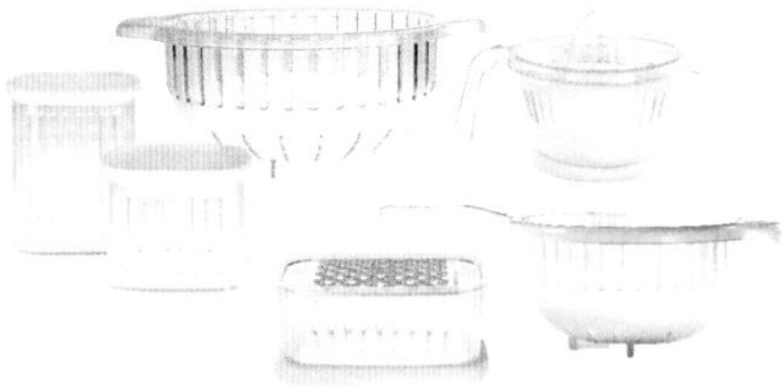

Bella Vista set designed by Sottsass Associati.

certainly has enjoyed a great revival, also because new plastics have been invented, more resistant and in a certain way less repulsive, that give more of an idea of solidity, substance, of durability over time as well.
The Guzzini company in particular has almost a century of experience in the field of plastic materials. Guzzini staff work constantly to keep abreast of the problems raised by plastics and the types available.
I am very interested in this job, because I'm curious, mainly to see how it all ends.
Ceramics and porcelain especially are materials that can resemble plastic, and vice versa, plastic can resemble porcelain.
Porcelain has an advantage, at least in my view. It's got weight, while plastic doesn't. And in my opinion weight is an important element when designing objects.
An object can weigh too much or not enough; household articles in particular, we hold them, position them, if we touch them they're not supposed to tip up, etc. There is a whole psychic and cultural relationship in the weight of objects. Ceramics has this weight while plastic often doesn't, so in this sense as well, plastic gives the impression of being temporary.
Ceramics gives the sensation of something old, as it were. There is something, at least for my generation, that evokes archaic sensorial memories. But I don't know plastic, I'm not sure, in short I still haven't "gone to bed" with plastic...
When we started out, we agreed that, at least in design, we should try and overcome this "fate" imposed by plastic objects. But there was the risk of returning to shapes or styles of carved wood rather than to antiquity. As a rule this was to be avoided.
Now we're working on something that doesn't necessary immediately convey the idea that it's made from plastic. In fact, when the objects are arranged on the table, in the kitchen drawer or picked up, they feel like real objects, not something to be used and disposed of.
I personally think that, at this time of total destruction of the sense of existence by unbridled consumerism and the speed of consumption, it's important to rediscover the awareness that an object can be a tool of life, not just something you use then throw away.
When I say tool of life, I mean something that really accompanies life; life is not only made up of gestures, but emotions, thoughts, nostalgia and memories as well.

June 1998.

Jean Michel Wilmotte

"In actual fact, I became acquainted with the iGuzzini company because I was a big user of iGuzzini products. I believe that I was the biggest user of Cestello modules for a number of years, as well as of other products. On several occasions I even happened to request specific models as it were, longer or wider ones, with several modifications, or different colour variations. This way I was able to get to know the people representing iGuzzini better and better; so eventually I began to say that I would like special modules, specific light fixtures for a particular museum, for offices. iGuzzini progressively commenced crafting products that were virtually made to measure, on my order, for my work sites. So I started designing products for them, because they were either ones that I didn't have or were lacking from the market, but that were needed. I therefore set off quite early to Recanati, where I visited all the plants and became convinced that it was a company that worked well, that was committed, located in a most pleasant area for work. At the same time, however, the facility was equipped with an authentic lighting engineering workshop, with people who were really dedicated! This was a serious business, one that drew me in. So we began working on a few products together.
But it took, it takes, time, because our relations go back five or six years now, so we have been trying to meet up and design together for at least five or six years.
This is briefly how it all began. Well actually, first of all, to get to Recanati I felt as though I was going to the ends of the earth, because I had to take a flight first to Milan or Rome, then another one to Falconara, then from there to Recanati. I really felt as though I was on an expedition. But then I realised that it was for a good cause, because instead of ending up in the suburbs around Turin or Milan, we would go to a place that was relaxing, where life was gentle.

Hublot wall luminaire, designed by Jean Michel Wilmotte in 1999.

When visiting the facility, I realised that it was a factory where people felt motivated to work – whether in research or production, but also in achieving quality standards in finishing processes, in product detail. Something that I really appreciated.
As regards Sirrah, it's a company that I've known for years, in fact it was probably the first lighting company that I ever worked for, when I left school; that must have been between the seventies and the eighties, when jersey was still in fashion.
It was very interesting, one of the first products that I had ever used. So when I heard that Adolfo Guzzini had taken Sirrah over, I was pleased, I thought "that's wonderful, now the company will get a new lease of life, and they'll find new products for their range"".

March 2001.

Eternal principles, today's values

Speech delivered by Adolfo Guzzini during the inauguration of the Galleria Civica Guzzini, Recanati, on the occasion of the exhibition *"Il Bello e l'Utile"*, June 1998

The exhibition "Il bello e l'utile" held in the Galleria Civica Guzzini, Recanati (1998).

The first generation of the Guzzini family initiated a process that time has shown to be successful: a gradual transition from the small business to industry, with a handful of clear but solid principles to guide the way: commitment to work; quality; love of one's own lands; constant attention to trends in tastes and lifestyles; the pursuit of innovation.

We, the entrepreneurs of today, share the principles that guided our predecessors, but besides them, we must observe other values and convictions. For example, the conviction that a company is strong when it belongs to a strong system; this is why we fervently believe that, in today's society, a creative, truly active relationship between enterprise and school, cultural institution and public administration, is vital. Further to generating revenue, an enterprise plays a decisive role in the development of the surrounding territory.

According to peasant culture, an enterprise belongs to the owner. But it would be more correct to say that it belongs to the owner if he or she can ensure its progress. The enterprise also belongs to the people who work in it – as long as they believe in it – and to the territory, but only if the awareness that enterprises are decisive for the development of the collectivity is asserted.

One of the ways that a company can contribute to the growth of the territory in which it operates is to provide for constant on-the-job training of its personnel and of those working in related enterprises. The major Guzzini companies dedicate between six and seven thousand hours every year to training schemes.

Training also extends to developing a specialised managerial culture, required by the entrepreneur for decision-making, given the extreme complexity of the system in which he or she operates. Corporate management today is characterised by complex issues. Technological problems can be complex: a press used for processing horn lasted decades without replacement; modern equipment can become obsolete in a matter of years. Markets are also complex: yesterday they were closed, competition scarce; today they are open, while competition is rife.

For all these reasons, professional managers have been working with the family, such as Antonio Santi and Antonio Renzi, the first to take part in this story. Family members are in turn asked to operate with a managerial culture, besides their entrepreneurial commitment.

Natural talent, curiosity, ethics, intellectual abilities are no longer sufficient for today's entrepreneur. He needs to be like Ulysses: if the will to act is there, the challenge is worth taking up.

The challenge of globalisation

An enterprise can make its presence felt in the global market thanks to the strength of the national system to which it belongs.

Foreign countries are not to be considered solely and exclusively opportunities for market expansion or production cost reduction, but rather as possibilities for cultural exchange, to the constant enrichment of enterprises and therefore the system in which they operate.

These exchanges generate a process of imitation that in some cases can be particularly profitable: many technological innovations have been brought to Italy from the United States or Japan, or from other European countries, such as Germany.

However the typical Italian approach to design, combining humanistic and technical culture, has come up with another innovation. In this global market, Guzzini companies have elaborated their own entrepreneurial model that could be described as that of the "pocket multinationals", in which the branches are vital junctions, generating a network of relations with other companies, with the opinion leaders of professional and cultural activities of the host country. Operating in the global market is not only about sales, but about culture as well.

The role of design

The American economist John Kenneth Galbraith wrote in 1997 about Italy, "design is something invisible, a heritage that cannot be shared out, assuming artistic and artisan resources that are acquired over many generations [...] It will be a strength for the Sistema Italia if you continue to translate it [...] into quality industrial production, updating technologies together".

Design plays a role of synthesis, as it represents the fusion of humanistic and cultural cultures, representing in turn the enter-

prise culture. The design of a product is not exhausted in its static (materials and components) or dynamic (production processes, technology) structure, but extends beyond services, communication, sales organisation – all facets without a physical configuration.
Design tends to integrate bonds, opportunities, possibilities: the designer, when he or she configures an article, contributes to the creation of the world's form, interpreting and proposing a material culture, rather than acting as a pure "beautician" of instrumental performance.
Design has now become a synonym of a conscious act of conception, by which the only development possible – sustainable development – can be achieved. In which the only demands to be observed are those of the quality of human life, of the environment, of work, through the pursuit of ergonomic and biocompatible solutions.
This activity can only be conducted by means of collaboration with important international research centres, with universities, and particularly those in the Marche Region. Further confirmation that a company can only be as strong as the system in which it operates.
Combining the Utile and the Bello, the Useful and the Beautiful, means to combine the skills of the left and the right side of the human brain in a single project – and reminds us all that the economy is part of culture, and not the other way round. This has always been part of the Guzzini group ethos.

Printed in Italy, 2002